REMOTE ACCESS ESSENTIALS

REMOTE ACCESS ESSENTIALS

Margaret Robbins

AP PROFESSIONAL

Boston San Diego New York
London Sydney Tokyo Toronto

This book is printed on acid-free paper. ♾

AP PROFESSIONAL
1300 Boylston St., Chestnut Hill, MA 02167

An Imprint of ACADEMIC PRESS, INC.
A Division of HARCOURT BRACE & COMPANY

United Kingdom Edition published by
ACADEMIC PRESS LIMITED
24–28 Oval Road, London NW1 7DX

ISBN 0-12-691410-9

Printed in the United States of America
95 96 97 98 99 IP 9 8 7 6 5 4 3 2 1

CONTENTS

ABOUT THE AUTHOR

Margaret Robbins is a computer programmer with over ten years' experience programming on UNIX and Windows systems. She is currently living in Europe, working as a programmer and computer journalist. Her previous publications include three other books in the Networking Essentials series: *Network Design Essentials, E-Mail Essentials,* and *Internet Access Essentials,* all with Ed Tittel; *AppWare: A Developer's Guide,* with Roy Christmann; and *Networking with Windows for Workgroups* and *Novell's Guide to NetWare 3.12 Networks* as part of Currid & Co., with Cheryl Currid. She has a degree in computer science from Rice University.

Acknowledgments

Any project of this scope involves many people. Many thanks to everyone at AP PROFESSIONAL, especially to Editor Jenifer Niles, Editorial Assistant Jacquelyn Young, and Production Manager Karen Pratt. The Waterside agency team of Carole McClendon and Belinda Catalona are always supportive and helpful, and are responsible for getting this series off the ground in the first place. For graphic design, typesetting, and manuscript polishing, Susan Price consistently does outstanding work. Dawn Jacobs' attention to detail was invaluable in making sure the final product was free of glitches.

On the content side, the help and advice of Roy Christmann, Alain Nimri, and Jens Wehde were especially important. Thanks also go to the thousands of software developers who are committed to making remote access available to everyone, simply and intuitively.

For moral support and patience, Volker Wehde is again unequaled.

Finally, without Ed Tittel's constant support, encouragement, and vision, not to mention wit and resourcefulness, none of this project would ever have happened at all. Thanks again, Ed— it's been a great experience with a thousand benefits.

INTRODUCTION

INTRODUCING THE ESSENTIALS SERIES

Welcome to the *Networking Essentials* series. This series aims to fill the gap between theory and practice for networking. It avoids jargon and unnecessary detail, and focuses on the concepts, topics, and techniques you need to simply accomplish your goals. It's aimed at professionals who use the various aspects of networking and communications, but who may not be interested in the technical underpinnings of the tools they work with.

This is the fifth book in the series. The first three, by Ed Tittel and Margaret Robbins, are *Network Design Essentials, E-Mail Essentials,* and *Internet Access Essentials.* The fourth book in the series, by Ed Tittel, is entitled *ISDN Essentials.* All are published by AP PROFESSIONAL.

About this Book

Remote Access Essentials is a guide to connecting to another computer in order to get work done. It tells you what you can do when connecting to other computers, and what you need to know to accomplish those tasks.

Since configurations and platforms differ so dramatically, this book orients itself more toward helping you find the resources you need than giving you detailed instructions that walk you step by step through the configuration and use of your remote access solutions. You could use this book quite profitably in conjunction with one of these step-by-step guides, but this book's focus is more on helping users of a wide array of different computer platforms. Therefore, instead of tutorial instructions, it gives you the background you need to ask the right questions, get the right equipment, and keep it running.

Because so much of communications, including phone call rates, regulations on purchasing communications equipment, and availability of digital or wireless options, depends on where you are in the world, the book points out the areas that differ across international borders, and explains where it assumes you are. For the most part it assumes you are in North America for the purposes of recommending sources, services, and companies, but where possible also gives access information so you can contact these sources from elsewhere. Furthermore, when explaining travel considerations, it gives you the information you need when traveling outside the United States to some of the most common business destinations, but the information could easily be applied in the opposite direction.

How This Book Is Organized

Remote Access Essentials is split into five parts. In the first part, Introducing Remote Access, you'll find an overview of the possible uses of remote access, and chapters presenting the terminology, concepts, and necessary technological information, from modems to wireless communications to ISDN to software.

In Part II, Applying Remote Access, you'll find wide coverage of the applications currently available for remote access users. From software and data retrieval to remote computer control to sophisticated messaging technologies to the myriad uses of the Internet, the information in this section will explain what types of things you can do with remote access, and how you might benefit from them.

Part III, Implementing Remote Access, takes the discussion to a concrete level. It summarizes the software and hardware considerations you need to take into account, depending on the remote access applications you decided you need in Part II. It explains how to evaluate your needs while planning for the future, offers tips on successful purchasing, and explains what you need to know to install the equipment. It then walks you through file accesses, Internet connections, and searching in some detail, to get you started, and offers lots of pointers on places to get more information.

The next part of the book, Part IV, Keeping It Useful, presents some strategies for using remote access in the long term. It starts off by summarizing all of the tips and trade-offs presented throughout the book, so that when you go to upgrade your technology, you can look for all of the information in one place. From there, it offers suggestions on finding an on-line service, computing while traveling, and telecommuting. It wraps up with a discussion of the security issues involved in remote connections.

The appendices at the back of the book offer two resources: a glossary that defines the terms in the book, including some that do not appear in the text but are important in using remote access; and a vendor list. This listing gives the names of products and manufacturers mentioned in the book, along with contact phone numbers.

About the Reader

This book assumes you know a little about computers, but not a great deal, and not at a technical level. You probably use computers in your work, but don't want to have to learn more than you have to in order to accomplish your tasks. You're interested in computers as tools, not as a hobby.

Therefore the book explains the necessary terms as they come up, but it doesn't go into the underlying technical details. It aims to give you just the amount of information you need to use the applications or features described. In some cases, the necessary detail is more than in others, because the technology is still not developed to a point to hide the details.

The book assumes that you need to be able to attach to another computer from a remote location to get work done. The emphasis here is on business-related purposes rather than hobby, so this book spends more time on financial databases than it does on games or chat services. The concepts are the same, though, so after you've downloaded those stock quotes and sent out the quarterly sales report to the field staff, if you want to settle down with a game that you downloaded from a remote site, you'll know how to accomplish that, too.

part I

Introducing Remote Access

How it All Works

Part I of this book serves as an introduction to remote access. Chapter 1 defines the terms you hear most often, and describes some of the ways you can apply remote connections. By the end of this chapter you'll know a great many things you can do with remote access, and you will have begun to get familiar with the language and terminology of remote communications.

Chapter 2 begins laying the technical groundwork, in high level terms. It describes the basic types of connections, and the components that all remote connections have in common. By the end of this chapter, you'll have a sense of what types of remote access are most suitable to various applications, and what the relative costs might be.

Chapter 3 presents the most prevalent form of remote access, through a wired telephone modem. In this chapter you get an overview how modems communicate, what types and speeds of modems are available now, and what all the letters, numbers, and abbreviations of modern modem technology mean.

Modems are not the only devices you can use to make remote connections, though, and Chapter 4 presents some of the alternatives, from cellular connections to high-powered digital lines. This chapter also takes a look at the highly portable *personal digital assistants* (PDAs).

Chapter 5 introduces you to the software you need to make remote connections work. In a quick survey, it discusses classes, features, and issues of communications software. Later chapters examine software in more detail with respect to specific applications; this chapter provides the overview for those discussions.

By the end of this part of the book, you will have a grounding in the basic ingredients of remote access technology, hardware, software, and connection media. The next part of the book builds on that by exploring in detail the things you can do over remote connections.

chapter 1

Why Remote Access?

Going Remote

The fact that you picked up this book shows that you already know that you want to connect remotely to another computer, but you may not know how to go about it or exactly what you want to do with the capability. This book will help you understand what you can do, clarify what you need and want to do, and set up everything necessary to accomplish it. This chapter introduces the possibilities and options available when you move to the remote access world. From expanded work options to simplifying your personal life to pure fun, the remote world as it exists right now has lots to offer.

Terms and Definitions

To set the stage for the discussions in the rest of the book, here are some of the basic terms and definitions that you'll encounter in the coming chapters. Most of them you are probably already familiar with; some of them have usages specific to the context of remote communications.

Remote Access

Remote access can mean any number of things, depending on your point of view. If you're a telecommuter who routinely connects your home computer to the office LAN, you may think of your home computer as an extension of the office environment, and not as remotely connected at all.

In this book the term is used in the broadest sense, meaning any means or mechanism you might use to connect to another computer and get things done. You might accomplish this connection with another computer, a fax machine, a pager, a PDA, or some other device. This book discusses all of these possibilities.

Desktop Computer, PC, Home Computer

These terms are often used interchangeably. In the context of remote computing, they all refer to the same computer: the one in front of you, wherever you are working. You might also hear this computer called a *local computer*, *local PC*, or *client PC*.

Host Computer, Remote Computer

This is the computer to which you are connecting. It is the one that offers some service or resource that you want to take advantage of; you are using your local computer to access that resource. When you look at it this way, it's easy to see why this one is called the host, and your local computer is considered to be the client.

Modem, Fax/Modem

A *modem* is a device you can attach to your computer (or often, that comes integrated with the computer) that lets you send computer signals over a phone line. A *fax/modem* is a modem with the added ability to send faxes—a related but distinct operation. Chapter 3 describes both operations, and describes how both devices work in more detail.

Link, Connection

These are more terms that can have several meanings depending on the context. They refer to the medium over which you transmit data. It can mean the active, current link that you are using, or the general link available to you over a longer period of time, whether it is active or not. The latter, though, is more often called an account.

Account

Some computers allow only one person to work on them at a time—these are single-user computers. Others allow more than one; they are called multiuser computers. Before you can connect to a remote multiuser computer, you almost always need an *account* on it. This administrative and security mechanism lets the system operator of that remote machine keep track of who is working on the machine.

So how do you get an account on that remote computer? Sometimes you can set one up entirely from your home computer, but often you have to contact someone in the real world—that is, by phone, fax, or postal mail. Once you have done that and made any necessary payment arrangements, the administrators at the remote site set up an account for you. At that point you can dial into that remote computer with your modem, tell it your account ID (also known as a user ID) and password, and then access the resources on that machine.

Logging On, Logging Off

Logging on is the act of connecting to a remote resource, and can include dialing to the remote machine (or establishing a link by some other means), entering your account ID and password, and initializing your remote session. The time during which you are working on that remote computer via a link is the time that you are logged on to that computer. When you have finished with your task and want to disconnect from the remote computer, in whatever way peculiar to that system, you are *logging off*.

Upload, Download

Think of your local computer at the base of things, and all the remote computers in the world as up in the sky somewhere. Thus, any time you send a file to another computer, you are *uploading* it, and the act of bringing a file, program, or whatever from a remote computer to your home computer is called *downloading*.

Uploading and downloading refer, by and large, to the transmission of large files or software packages. The vocabulary for the more basic remote operations, such as e-mail and fax, reflect a more postal-oriented model: you send and receive them.

E-Mail

Giving you the means to send and receive electronic messages, *e-mail* is one of the most widely used applications of remote access. A few years ago, if you had e-mail at all, there was a good chance it was only at work, and it only connected you to people at work. With interoperability greatly improved, e-mail is now nearly as ubiquitous as fax, and many people have their e-mail addresses on their business cards. Chapter 6 looks more closely at this phenomenon.

Interactive Connections

An *interactive connection* is one in which two or more people are connected to a common computer at the same time. When one person issues a message, the other person sees it immediately and can answer right then. This distinguishes interactive connections from another type, *asynchronous connections,* in which you issue a message and the receiver retrieves and responds to it at some later time. You can think of it as the difference between making a phone call and leaving voice mail.

Interactive connections are not the most common mode of remote access for two reasons: people like the convenience of being able to respond at a convenient time rather than when the message comes in, and more important, maintaining a live link can be expensive. Interactive links may become more common, though, as technology improves and costs come down. They're essential for interactive games, chatting, and video conversations.

For simple conversations, however, telephones remain the most convenient option because of their wider availability.

Other Terms

A number of other terms appear when discussing remote computing, and the book will define them as needed. The ones defined above, though, come up frequently. If you run across other unfamiliar terms while reading the book, or familiar terms that are used in unfamiliar ways, check the Glossary. Many more are defined there.

What You Can Do with Remote Access

Now that you have a grounding in the language of remote access, let's look at a few of the many applications of remote access. This section doesn't cover the items exhaustively, it will only to give you an idea what the possibilities are. Later chapters will describe each of the applications in much more detail, from what you can do with them to how to make them work. This section serves as an appetizer.

The Basics: Faxes and E-Mail

Most people get their first experience in remote access with these two items. Since faxing at its most basic does not require any computers at all, you might wonder whether it really belongs in a book on remote access. But since most modems today, and most of the available connection software packages, handle faxing as well as computer communications, and since faxing is so often confused with remote communications, it is included here in order to describe the distinctions.

You have probably sent faxes before. Doing so usually involves feeding sheets of paper into a machine, which scans the lines on the paper and translates them into a digital representation, then sends that representation over the phone line. The receiving fax machine must operate on the same format; it can translate the received pulses back into a graphical representation of the original document. It then optionally prints that image onto another piece of paper.

Since a significant part of this scenario involves an electronic form of the document, it's possible to bypass the paper aspect of the operation, and use your computer to compose an electronic document, convert it to the electronic format understood by fax machines, and send it using a fax/modem. Similarly, you can receive faxes with your fax/modem that come up on your computer screen, and store them on your hard disk, without ever having to print them. The capability to send and receive faxes on your computer requires both software and hardware; both now come with most modems.

E-mail is quite a different beast, but it is now almost as ubiquitous as fax capabilities. Most people first encounter e-mail at work, where it enables them to communicate with colleagues in other buildings or states, as well as those right down the hall. In the workplace it can save a lot of time and paper—you can use it to make announcements or deliver memos that would once have been printed out in batches of hundreds, then thrown away immediately after being read.

Being able to read their workplace e-mail from home is often the first foray many people make into remote computing. This enables them to handle questions, send information or reports, and issue memos, all without setting foot in the office. This capability is all some people need to allow them to work from home.

E-mail is more and more important for personal communications, too. People use it to communicate with friends and family, across town or in another country. Chapter 6 discusses e-mail and its messaging cousins in more detail.

Intermediate Remote Access

As you move a step up from sending faxes and e-mail, the world of remote communications branches out dramatically. With a bit more effort, you can send computer files, acquire the latest releases of some software packages, or listen to the new efforts of a favorite band.

The distinction between e-mail and these capabilities is really rather artificial, of course. E-mail and fax are somewhat simpler because they've been around long enough for the software interfaces to have evolved to the point that they're now fairly easy to use. Intermediate remote access applications in this section are somewhat less standardized, and so will require a bit more work and information to make them useful. Some of them, to be truly useful, may also require more elaborate equipment (at least a faster modem) than is necessary for e-mail.

Much of this arena involves getting information, but the term "information" is deceptively vague. It could be a bug fix to a software package that you bought (or the software package itself), quotes on stocks that you have been tracking, research data on a company you want to work for, or the current weather forecast for your area. Some recording companies are now making short clips of their bands' newest releases available; you can download them from their computers, and with the right hardware and software, play the clips on your own computer before you decide to buy the CD. You can subscribe to electronic clipping services that will send you recently published articles that relate to topics of your choice. In some areas, it's possible to connect a satellite receiver to your notebook and use it to communicate with GPS (Global Positioning System) satellites, tracking your position, destination, and route on a continuously updated map.

Other possibilities include conversations with other people in on-line forums, also called newsgroups; buying and selling stock; making banking transactions, including paying bills and checking account balances; making travel arrangements. At the most interactive level, you can attach to a remote computer and run programs on it as if you were right in front of it; if your work requires software that is only on that computer, this means you can now work from home—or wherever you happen to be.

Fancy Interactions

These applications are considered more advanced uses of remote access not because they're necessarily more complicated (though they may be), but because they involve interactive connections, which tend to be more expensive than the asynchronous variety required for most of the applications just described.

In this category you find interactive games, where every player can see and react to the others' actions in real time. You also find chat connections, where two or more people hold on-line conversations via their keyboards.

It's also possible to hold voice conversations over computer networks. Since the telephone system seems to be pretty efficient, you might wonder when this would be useful. You are in fact better off using the telephone system for local calls in the United States, which are not metered. But if you're calling internationally, for instance, the cost of the network link can be orders of magnitude lower than the cost of an internationally dialed call.

On the far end of this scale you find interactive design, video, and other high-bandwidth, real-time applications. While it's possible to implement these types of applications over ordinary phone lines, they still need high-capacity digital lines for acceptable performance. These lines can be expensive, even for short periods of time, and the equipment needed to connect to them adds another significant cost to this type of connection. The benefit of being able to see as well as hear the participants in the call, however, is in some cases considered worth the extra cost—though at present you find these types of systems only in large offices, seldom in the home environment.

Summary

Now you have a clearer idea of the possibilities available to you through remote access, and you have been introduced to the terms and definitions that you need to talk about it. The next chapter gives you some background in remote connections, including a broad overview of how the technology works.

chapter 2

Communications ABCs

A Look at Communications

Recently—in the last 15 years or so—two things have been happening that make remote access possible. Computers and computer components have been at once getting much faster and much cheaper. At the same time, computer networks have grown dramatically more interconnected. Not long ago, a network meant ten or so computers that could communicate with each other, but now, many LANs of that size are connected to larger LANs, which in turn may be connected to even larger LANs or WANs. The result is that a large portion of networked computers in the United States are able to communicate with each other, at least on the relatively rudimentary level of e-mail.

Like so many aspects of the fast-growing computer industry, this growth has not been particularly orderly or easy on the users (nor on the manufacturers). Because things grew from so many directions at once, there are now a great number of different ways of doing the same thing. While the chaos is now quieting down, and in many areas industry-wide standards are emerging, the field is still filled with entrants, and finding your way through to what you want to accomplish can be confusing.

This chapter is designed to tone down the cacophony and present the offerings in a way that shows you their overall context and how they relate to each other. It starts by discussing at a conceptual level what all types of remote access have in common, then describes many different specific types of remote access in terms of these common features.

The Parts of Remote Access

All types of remote connections have four components: the medium over which the connection runs, connection equipment to act as an interface between the medium and your computer, communications software to manage the connection, and a language or protocol that specifies how it all works.

The Medium

The array of remote connection media is vast and growing. While a few years ago you had little choice outside the basic analog phone line, now you can use the cellular phone network or satellites to transmit your data.

Analog Lines

When you pick up the phone and call someone, you are making an analog call. This is true even if you are using your cellular phone, or your briefcase-sized satellite transmitter phone. The call is an analog one in any case, because the sounds you make when speaking are transmitted as analog signals—waveforms. You might lose some of the detail of the voice you're talking to, but the human ear is highly adaptable and can easily compensate for such losses, and can listen around clicks, pops, and static that can occur on the line.

Digital Connections

Digital lines behave rather differently. Rather than sending voice information over the line in waves, they use on-and-off signals (hence the name). You can still run voice phone calls over digital lines—in fact, these lines are often of better quality, with less static and other noise—but where digital lines really shine is in the area of computer communications. Those clicks, pops, and noises on analog lines that your ear might barely notice can be very hard for a computer to deal with; digital lines usually offer a much cleaner connection.

There's a perception that in order to run a digital line, you need fiber optic cabling, but this is not true. The difference is in the way the connection equipment sends the signals over the line (as described in the next section and Chapter 4). It's quite possible to run digital connections over copper wires; it is done all the time. What fiber does offer, though, is the ability to transfer a much larger volume of data.

ISDN (Integrated Services Digital Network) is the best-known form of digital lines in the United States, and it looks like its long promise is finally becoming reality. It is now possible, after years of anticipation, to get ISDN lines in your home. (If you don't yet see any reason to do so, later chapters might entice you.)

These are not the only types of media, just the most often used. Those who regularly need to send very large quantities of data can arrange for dedicated lines, which typically have fixed endpoints. That is, when you pick up the phone and call someone, you can pretty much call anyone else who has a phone; when you make a connection over a dedicated line, though, you always connect to the same site.

Wireless Connections

You don't find wireless connections in as many different settings as you do phone-type connections, but that is changing fast. In this category infrared, cellular/radio, and satellite connections are all vying for your attention. Infrared technology is already familiar to you; your TV remote control uses the same mechanism. It has the smallest range of the three, since the two connected devices must be in the same room. The next notch up the scale is the range offered by cellular telephone access. As long as you are within one of your network's cells, you can use this technology to keep yourself remotely connected from anywhere you can use a cellular phone.

Farther afield (that is, a range not limited to the reach of your cellular phone network provider) is satellite transmission. It's now possible to get a satellite telephone the size of a briefcase and to access via satellite any two points on the globe. Handheld versions are on the way. The limiting factor here is the network of satellites that receive and transmit the signals. The current network can link these larger transmitters together, because their signals are strong enough to reach these higher satellites. For the smaller transmitters to work, we'll need a new set of satellites that have somewhat lower orbits to reach the weaker signals. This won't be cheap, but a group of companies is already looking at the possibilities.

The Equipment

The equipment required to make the connection depends on the medium over which you want to run the connection. To connect to another computer over an analog line, for instance, you need a device that can convert your computer's on-and-off (digital) signals to the analog waveform that these lines can carry in a process called *modulation*, and the receiving end also needs such a device to translate (*demodulate*) the incoming waves back into digital format. The device that performs these operations is called a modulator-demodulator, or modem.

When your computer connection runs over a digital line, you don't need a modem anymore—your computer can issue signals in its native digital format, the digital lines carry them unchanged, and everything is still in digital format when it arrives at the receiving end. What you do need, though, is a device called a *terminal adapter*, which converts the digital format from your computer into the packet format used by your digital lines. If you want to place a voice call over the digital line, though, then you do need a device like a modem that converts your voice signal into digital format before sending it out over the line. This capability is built into ISDN telephones.

Transmission over dedicated lines can be more complicated, because the company that provides you with the line may use its own private format for sending data over that line. If this is the case, you'll have to get connection equipment from them as well.

Infrared connection equipment consists of a device that can emit signals in the infrared band of the wave spectrum, and a device that can receive and interpret them. This equipment is as direction-dependent as your TV remote control, and as unforgiving of obstacles such as furniture or walls, and has about the same range of use. Thus it's only useful if the two computers are in the same room.

To make cellular connections, you can get a gadget that attaches to your cellular phone and plugs into your computer. It works much like a modem, translating your computer's digital signals into an analog format that your cellular phone can understand. The same principle is at work when you make satellite connections—your transmitter in that case is sending the signals to an overhead satellite rather than your cell transmission point.

The Software

Depending on the medium you're using for your connection, or more specifically, depending on your connection equipment, you may have to run two "layers" of software on your computer. The lower layer, if you need one, is a driver that handles all of the communication between your operating system and your connection device. The upper layer, the connection software (also called communications software), is the one you use to place a call, send a file, or retrieve messages that have been sent to you. In many cases, the two layers will be integrated into one piece of software and you won't have to know anything about the distinction.

You're likely to make more changes to the configuration at the level of the connection software. This is where you define whether you're sending a fax, a picture, a note, or a program file. You also specify here who you intend to call, and you may have to specify some information about the caller's configuration. Chapter 5 expands on the various aspects and features of connection software.

The Protocol

Clearly, in order for any two devices to work together, they must agree to transmit information in the same way. Otherwise, a fax machine could be sending bits in one order, but the receiving machine might expect them in another. The result would be chaos. But instead, fax machines agree on a common standard, and the result is a fax.

So it goes with all other types of remote access. In some cases, the common protocol is the one the equipment manufacturer dreamed up, and only its equipment and software offer it. This is what's known as a *proprietary protocol*. This can be great for the manufacturer if all the other vendors agree to use its mechanism; the manufacturer makes lots of money and gets to call its way of doing things "The Standard." But this can put limits on you, the consumer: If you have purchased equipment for two sites from this manufacturer and you want to buy parts for a third site, you have to buy from the same company again, and if you acquire equipment from another company, it might not work with your existing equipment. Worse, if you want to communicate with someone who has invested in another manufacturer's solution, you're out of luck. Eventually the market settles down enough so that a generally agreed-upon standard emerges, but the intervening period can be painful for consumers and producers alike.

The other way of establishing a common protocol is to have an impartial board of international experts agree on one. The International Telecommunications Union's technical branch (ITU-T, formerly known as the CCITT) performs that function for telecommunications matters. Based in Switzerland and directed by the United Nations, they study communications and technology, with the help of experts from the related industries, and issue documents that define how two systems should communicate. The standards are usually named after these documents; this book lists them where they apply.

The presence of standards in a particular area of computer communications can be a boon for the consumer, because manufacturers then aim to woo customers more on price than features—the feature set is more or less established by the standard. Not only does it mean that your equipment might become cheaper by competition, but also it can be easier to buy. All you have to know is what device you want and the relevant standard for its operation, without needing to know the details of what that standard means.

Each type of equipment you buy should conform to the established standard for that type of connection, but this doesn't always mean the latest standard is the best. The newest ones can change before they get widely established, resulting in "flavors" of standards (which are not really standards at all). Investing in one of these before it's firmly established could leave you with obsolete equipment that you can't connect to anyone else (as the owners of Beta format VCRs well know). On the other hand, investing in relatively old technology might mean purchasing a solution with much less power than you actually need. This book advises you on the most practical choices wherever they come up.

Some Common Connections

To help solidify these concepts, this section discusses some of the most common uses of remote access in terms of the media, equipment, and standards that they require, and compares them on the basis of cost.

Fax

Sending a fax is not strictly a means of remote access, but it's included here for several reasons. Many people new to remote communications confuse fax and modem communications in various ways. To add to the confusion, most modems now come with fax capability, which is very easy for the manufacturer since both types of transmission use analog phone lines. This section should help clear up the confusion.

Fax operates by reading a graphical image of a document, converting that image to a digital approximation (this process is called *scanning*), and sending that digital version of the image over the phone line (after modulating it, as described above). The fax machine at the other end must decode the incoming waves back into digital format in order to reconstruct the image into something that can be viewed. Most fax machines then print the image onto paper, but a fax/modem's receiving software can just store it in the computer, saving paper. To send images that aren't already in your computer, though, you'll need a scanner.

The standards for fax transmission are specified by the ITU-T, and the most common today are Group 3 and Group 4. Group 3 was the first to use a digital mechanism (Group 1 and 2 fax machines were entirely analog), and using that standard, a typical page of text can be transmitted in under a minute (compared to six minutes for Group 1 machines). Group 4 machines are designed to run over digital lines and are therefore more expensive and less common than the nearly universal Group 3 machines.

Modem Connections

Whether calling up a bulletin board in your neighborhood to discuss local politics, or calling into the office LAN, when you make a modem connection you are making the most economical and widely used type of remote connection. As long as the call is a local one in the United States, the call itself is essentially free. The equipment, too, can be very cheap—the most expensive, fastest, tiniest modems today go for well under $500, and less elegant ones won't even cost you $100.

ISDN Connections

When you hear politicians arguing for a ramp to the data superhighway in every home, this is what they mean. These are currently much more expensive than standard phone lines, and with their supercharged speed, are still appropriate only for very high-volume users. But as more and more applications come on line, and the greater coverage brings the costs down, you might find that you need an ISDN link too.

Infrared Connections

The transmitter and receiver hardware for these devices are relatively cheap (a couple of dollars in most cases), making them an attractive option for equipment manufacturers. There are hidden costs though, as with any wireless solution. One is batteries. Nickel-cadmium batteries are rechargeable, but tend to be heavy. The lighter chemical batteries have lower power and a nasty habit of turning up in landfills and poisoning the groundwater, despite toxic waste disposal centers. And all batteries tend to run out of power just when you most need them.

Other disadvantages of IR connections are their limited range and somewhat limited speed—considerably slower than digital lines but about the same as the top speed for an analog connection. Despite the drawbacks, IR connections can be perfect for such simple tasks as uploading the address file in your PDA to your desktop PC. They are also appearing in printers, so you can print a file from your laptop by pointing your computer at the printer and starting the print job.

Wireless Connections

When you make a local call in the United States, the call is essentially free. You pay the local telephone company a fixed price each month, and in return you get all the local calls you can use.

Long-distance calls, as well as most local calls outside the United States, are metered by the minute, so how much you pay depends on how long you stay on the line. The same is true for wireless calls, and this feature can make them much more expensive than their wired counterparts, aside from the higher cost of the connection equipment itself.

The convenience, however, is hard to beat. If you move around a lot, wireless communications free you from having to juggle phone cards, not to mention finding a phone that takes your card. This type of connection generally offers about the same transmission speed as a standard analog line; the big advantage here is mobility (which is, however, limited by the extent of your network).

SUMMARY

Now you have an overview of various types of connections, how they work, and what you need to implement them. You know what the different connection types are best suited for, the equipment and relative cost of the various solutions, and how they are most often used. Most important, you have some background in what standards apply to each type of connection, which will help you to ask the right questions when you go to purchase equipment. The next three chapters look at the mechanisms behind remote connections in detail.

Analog Access

c h a p t e r 3

Modems Demystified

In spite of the advances in the last few years in connection mechanisms, the majority of connections by far are still made through modems over analog phone lines. Since this is the type of connection you are most likely to use, at least to begin with, this chapter is devoted entirely to explaining how this type of connection works.

The Anatomy of a Modem Session

In case you're not already familiar with how to use a modem, and what happens at a very high level when you do, we'll go over the basic chronology, to give you a point of reference for the more technical explanations later in the chapter.

To initiate the session, you use your communications software to select the party or service to connect to, perhaps by entering a phone number or selecting the target from a list (in which case the phone number must already be configured in the software). The software then sends an initialization command to the modem, which the modem acknowledges. The software then sends a command directing the modem to take the phone off hook (not physically of course; this means electronically engaging the line). The modem takes phone off the hook and dials the number that the soft-

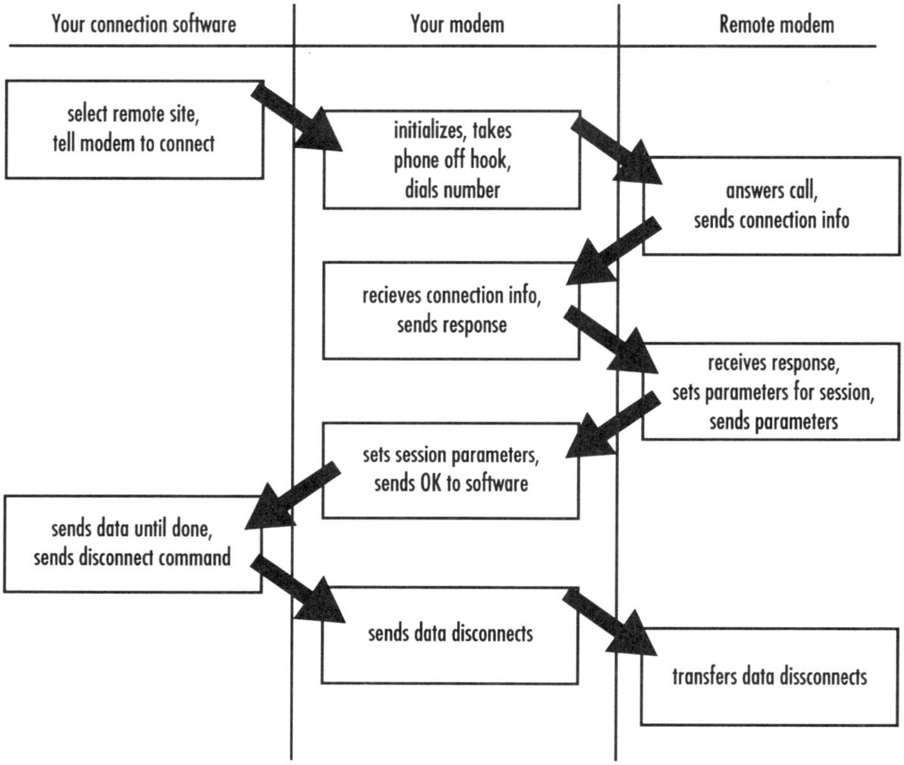

Figure 3.1 *Chronology of a typical modem session.*

ware provides. The modem at the other end answers the call, and the two exchange information on how fast they can operate and how they can communicate, and whether they can correct for errors. Once this is complete, the modem signals the communications software that it's ready to go. The rest of the session depends on what the communications software is trying to accomplish; it sends commands to send or retrieve information, or sends or receives data, until the task is complete. The software at one or the other end of the link terminates the session when the task is finished.

With that overview in mind, we can look more closely at the mechanisms and conventions that make this whole conversation possible.

BITS TO WAVES

You've probably seen diagrams showing so-called voice prints; they illustrate the waveform of the human voice by charting frequency against time. You can draw a similar diagram for any sound. The highs and lows indicate the rising and falling frequencies of the sound. Figure 3.2 shows an example waveform of a sound.

This is the type of signal that analog lines are designed to carry. Computers, though, have a hard time with these fuzzy, imprecise sorts of inputs. To convert from an analog signal into the crisp, on-or-off format that the computer likes, you need to *demodulate* the signal into a pulse-style signal that the computer can reliably measure. This is done by naming a cutoff point, and deciding that all waveforms that rise above the point represent ON pulses, and all below represent OFF. (The characteristic measured could be the amplitude, frequency, or phase of the wave, depending on the speed of the modem, but you are unlikely ever to need to worry about it.)

That's what the receiving side of the modem does. Naturally, the same thing is happening on the sending side, in reverse: it is viewing the data coming from the communications software, and wherever there is an ON bit, sending out a waveform above the level, and wherever it sees an OFF, sending a wave below the level. This side of the process is called *modulation.* The word *modem* is a shortened form of the term modulator/demod-

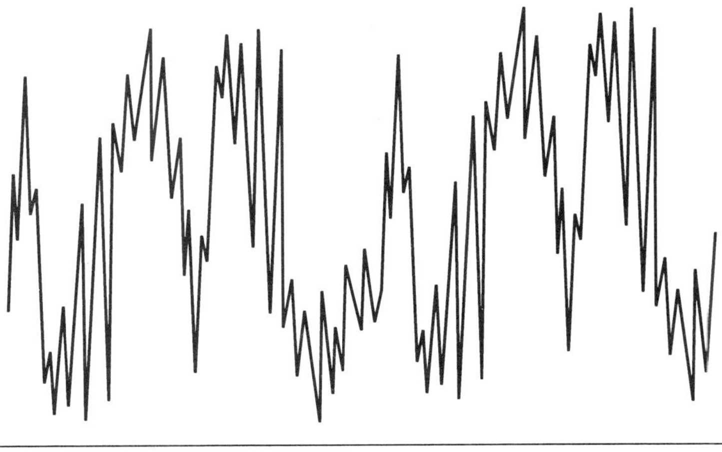

Figure 3.2 *Diagram of a sound wave.*

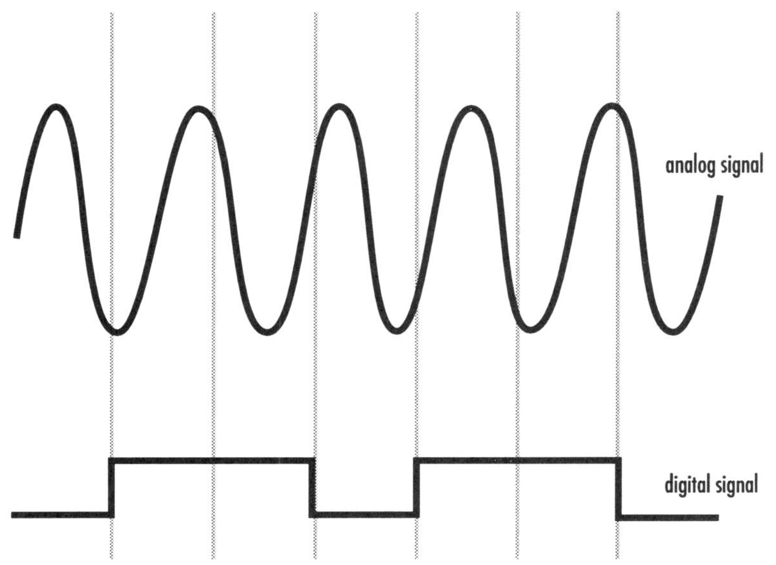

Figure 3.3 *An analog and corresponding digital signal.*

ulator. Figure 3.3 shows a digital pulse pattern, and the waveform that corresponds to it.

Fax machines operate on the same principle, but they use a different pattern than a modem for the bits in the waves. Therefore, you can't use an ordinary modem to send a fax to a fax machine, and you can't use a modem to receive a fax. Most modems today, though, come with circuitry and software that allow them to handle fax exchanges. These are called fax/modems, and they simplify the task of connecting to various types of equipment, while sometimes confusing the terminology.

ERRORS AND CORRECTING FOR THEM

The above discussion explains why you can use the same line for both the inherently analog signal of voice communications and the strictly digital version your computer requires. But once all those pulses arrive at the other end and are demodulated back into ones and zeros, the next question is how the remote modem (or yours, when you are on the receiving end) reassembles it all into something meaningful.

There are some obstacles. One problem is the unreliable nature of telephone lines. While perfectly fine for relaying voice, because the human ear is so flexible, the highly inflexible computer finds them considerably more troublesome. Imagine a spike in the waveform in Figure 3.3; it would be translated into an ON signal in the pulse form, destroying the integrity of the data being sent. Such spikes are not particularly uncommon on voice lines.

The simplest way to communicate would be to just send the ON or OFF bits across the wire and hope they get there. As the spike example demonstrates, though, the unreliable nature of telephone connections make this unfeasible. Some bits might get dropped, others in a burst of static might get their values switched to the opposite value. To compensate, engineers have developed a number of techniques to determine whether the bits that arrive at the receiving end are the same group that had been sent.

Here's how it works. The data being sent is broken down into bunches, called *words,* containing a certain number of bits (ranging between six and nine, but usually eight). Each word is preceded by a start bit, indicating that a new word is coming, and the end is marked by one or more stop bits. Sometimes, the whole word is followed by a parity bit—this is where the error detection starts.

The value of the parity bit depends on the number of ON bits among the data bits. The parity bit is set to ON or OFF according to whether there is an even or odd number of ON bits in the data packet. There are two types of parity, even and odd. If you are using even parity, for example, and there is an odd number of ON bits in the packet, the parity bit will be set to ON to keep the total number of ON bits even. When using odd parity, the goal is to keep the total number of ON bits an odd number, so the parity bit in that case would only be set if the number of ON data bits were even.

This way, if the receiving side is running even parity and gets a packet of eight bits with an odd number of ON bits, but the parity bit is off, it knows that an error occurred and can request retransmission of the packet, or signal an error to the communications software.

This is somewhat confusing, but unfortunately modern communication is still not automated to the point that you can ignore it altogether. Many times when you are given the access number for a computer host, the provider will also specify the number of data bits, stop bits, and parity type (if any). In most cases, though, you can just enter these values into your communications software and forget about them.

Heavy Error Correction

Parity checking only provides a means for detecting errors at a very low level, within every group of data bits, and no means to correct them. There are other algorithms that can not only detect errors, but correct them. Such algorithms may be built into your modem. In the best case, your modem can communicate with the other modem after establishing the connection and together they can determine what types of error correction they have in common, if any, and arrange to use one without getting you involved at all. The discussion on the Hayes command set later in this chapter explains why things might not always have this happy outcome.

Error correction can also happen at a much higher level, in the communications software or file transfer software. In fact, most file transfer protocols support some type of error correction.

Transmission Speeds

The single most important statistic about your modem is how fast it can run. This speed, also known as the data rate, is measured in *bps* (bits per second). A 2400 bps modem, a respectable speed only a few years ago, can send a 50-kilobyte file (about the size of this chapter, including formatting information) in around five minutes. A now-commonplace 14,400 bps modem can do it in less than 30 seconds. While the data rate is not the only thing you need to know when buying a modem, you can see why it deserves a lot of consideration.

The first devices that could send digital data over phone lines ran at a rate of 110 bps, in the late 1950s. By the mid 1970s, you could get modems that ran as fast as 300 bps, which was a big help. In the early eighties 2400 became the luxury speed, followed shortly by 9600 and 14,400, and now 28,800 bps modems are the new standard. Today a very affordable modem can handle 14,400 bps, automatically step down to a lower speed if the line is too noisy for the requested speed to work, compress data if the other modem agrees on the method, automatically detect and correct errors in transmission, and send faxes.

Most modems can negotiate a speed based on the noise on the line and the highest rate at which they can both operate. In a noisy environment, therefore, you might see much lower rates than you would expect, and on very noisy lines you may even have the connection drop altogether. This can be very aggravating when you are trying to transfer a large file. If it happens often, call the phone provider or the remote service provider. There may be a problem on the line or with the remote equipment.

How Fast Is That?

The terminology of transmission speeds is confused by a couple of things. First of all, an older term for measuring transmission speeds still crops up: *baud*. The baud rate is not the same as the bits per second. Rather, it measures the number of signal changes per second. In older modems, one signal change encoded one data bit, but the newer, faster modems can encode more than one data bit in each signal change. Thus, to be precise and consistent, use the bps rate rather than referring to the baud rate.

Another simple error is to miss the "k" that appears in the rate specifier. Older data rates were always measured in bits per second, or bps, but when rates crossed the 1000 bps threshold, it became common to refer to *kilobits* per second (kbps). Thus a 28,800 bps modem is also a 28.8 kbps modem. The latter form is used more often in this book.

Compression

The addition of compression to most newer modems is another confusing detail. Now that chips are so fast, they have time to examine the data stream as it goes by. If they see several instances of the same data value in a row, they can insert a code indicating the repeats and just indicate how many to reconstruct at the other end, rather than taking the time to send them all.

This is the simplest kind of compression, called *run-length encoding*. More sophisticated, and efficient, methods can detect and compress patterns in the data, and can compress at rates of up to 4:1, giving you an effective fourfold increase in speed—but only when communicating with another modem that runs the same type of compression. Note whether the manufacturer is reporting the effective transfer rate or the actual data rate before you buy. A 2400 bps modem with compression, for instance, could advertise an effective transfer rate of 9600 bps, but it's still just a 2400 bps modem if the modem at the other end doesn't share the convention.

Modems and Standards

One thing that made modem technology so slow to take off (from the 1950s to the middle 1980s) was the lack of standards. Now there are lots of standards, and they apply to both sides of the modem, the line side and the software side. Not only must the modem send data over the line in an agreed-upon way, it must handle errors in the same way as the other modem (as discussed above), and it's extremely helpful if they also converse with software in a uniform way. In fact, the adoption of a standard software command set (the Hayes command set) was a big factor in the rapid growth of modem communications. This section examines each of these types of standards.

Standard Ways of Talking to Each Other

There appear to be a great number of different standards floating about in the modem world today, and knowing which one you need can be baffling. The box the modem comes with is covered with acronyms and abbreviations, but it's really simpler than it seems.

It used to be easier—all you had to look at was the data rate (speed). Now, with data compression and error correction built into modems, it's important to know whether the compression or error-handling mechanisms match the ones everyone else is using. If not, you can only use the features when talking to another modem of the same type.

The older V.42 standard supports built-in error correction and data compression. It uses the MNP standard for error correction, while the modified V.42bis uses its own error-correction method. Either way, V.42 modems use an older, block-oriented mechanism for sending data; the newer V.34 modems use a streaming technique, which makes them much faster. While V.42 modems can operate at speeds up to 14.4 kbps, V.34 modems can run up to 28.8 kbps. In short, if you are buying a new modem today, look for V.34 support.

Standards for Talking to the Software

Modems not only need to talk to each other, they must be able to converse with the connection software. If all modems respond to the same commands from the software, you can use any software with any modem, but this requires an agreed-upon standard command set. The Hayes modem, first introduced in the early 1980s, quickly established its command set as that standard, meaning that nearly all modems now implement the same command set.

The Hayes command set specifies a set of commands that a software program can send to a modem, and responses that the software can expect to receive from the modem. Standardizing this mechanism greatly simplified both installing a modem and communications software, as well as the writing of communications programs. For this reason, it's both difficult and a poor idea to buy a modem that does not support the Hayes standard.

There's a catch, though. The problem is that the Hayes command set was developed more than 10 years ago, and was not designed to deal with all the features modems have today, such as data compression, automatic rate changes, high speeds (4800 bps was the original top speed), and error correction. The command set has been expanded by a number of manufacturers, including Hayes, but there is no definitive standard for the expansion commands. The situation is improving, but you should know that you may have to tweak your communications software and get to know your modem's command set in order to use all of its features.

Modem Commands

The Hayes command set consists of 2-, 3-, or 4-character commands that your communications software sends to your modem. **AT,** for instance, stands for ATtention, and acts as prefix for most commands. The sequence **+++,** as another example, tells the modem to disconnect from the link. The single letter **A** instructs the modem to answer a call.

Clearly, you would not want to have to enter these commands by hand during a communication session (though some people do). Instead, the communications software automates all of this for you, but you may still have to specify some things by hand before you start. The software has an entry, for instance, where you can specify the initialization string—this is the set of commands the software sends the modem at the start of a connection that tells it whether to use compression, error correction, and other features.

Command Settings

As explained earlier, the places you're most likely to run into compatibility problems are in the commands that turn on the advanced features (including the high speeds). Generally, you arrange an initialization string that includes all of the fancy features you want to use; if the other modem can't handle them (or recognize them), it ignores them. You may have to look in your modem's manual, though, to find out what they are.

Another place to look for these commands might be in the software that comes with the modem, but this is not as promising for two reasons. First of all, you may have to wade through several unfamiliar screens to find where it stores the string you need. If you plan to use that software for all of your connections anyway, this is not a wasted exercise, but if Chapter 5 convinces you that you need higher-powered software, it might seem like one. The second reason is more important, though; it's possible that the software itself has not been initialized with the correct commands either, especially if it was written to work with many different brands of modems. Either way, you're likely to have to open that manual at some point, if you want to ensure that you're taking advantage of all the features.

SUMMARY

You now have a good idea how an analog connection works, and what can go wrong. It sends and receives data over an analog phone line in a certain format, though there are variations in that format that you need to watch out for. You also now have a useful overview of the Hayes command set, which will help you more quickly configure things when you install your modem and communications software.

chapter 4

POWER AND PORTABILITY

BEYOND THE MODEM

The last chapter discussed the simplest and most common remote connection mechanism: using a modem over an analog phone line. It also discussed the modem's limitations, including a current top speed of 28.8 kbps and relatively non-portable package (because of the wires and power requirements).

Millions of users are perfectly happy with 2400 bps data rates, and if you are sending text messages and occasional files, you may be too. But for data-intensive applications, even 28.8 kbps can seem limited.

What's needed at that point is a way to boost the data speed even higher. More speed means more data goes over the same size link. In discussing the ways to make that happen, two words come up a lot: *bandwidth,* which refers to the potential carrying capacity of a link, and *throughput,* a measurement of the actual amount of data transferred at a particular time. If data links were water pipes, bandwidth would refer to the size of the pipe and throughput to the volume of water transmitted at a particular instant.

This chapter describes the options available to get past both the speed and the portability barriers of the modem. It discusses using digital lines to increase your bandwidth well beyond the 28.8 kbps analog limit, and it describes the various wireless technologies and portable computing options to free you from the geographical constraints.

High-Powered Links

One of the best things about analog lines is their sheer ubiquity. They reach almost everywhere in the world now, so you can run computer connections between nearly any two points in the world for the cost of a phone call.

Recall from the previous chapter, though, how much work is involved in translating your computer's language of bits and bytes into analog signals that the phone lines can handle, and translating those waves back to a digital format at the other end of the connection. If you could eliminate all that work, the communications devices could run much faster. Digital lines do exactly that.

How Digital Lines Work

As described before, digital connections send bits over the wires rather than waves. They behave much more like your office LAN than like telephone lines. In fact, the implementation of digital transmission is based on a networking model, and you'll hear providers of digital lines called *network providers,* and the collective sum of all digital lines is often called *the network* by people who work with them often.

Digital transmissions can run over fiber, copper, or other media. It's a misperception that all digital lines require fiber optic cables, at least at the low end of the scale. As things get faster, though (above about 128 kbps), copper does reach the limit of the signaling rates it can handle. Thus, even if your area does not have fiber optic lines running to every house, you still may be able to get low-speed digital service, but you will almost certainly need fiber wiring for the higher data rates.

The factor that determines whether you can get digital service is not necessarily the wiring, but whether your local service provider has installed digital switches in your area, which operate on very different principles than analog telephone switching.

ISDN

It's difficult to talk about digital lines without talking about ISDN. This stands for Integrated Services Digital Network, and it is by far the most widespread standard for digital connections. Today, if you have a digital line anywhere in the world, chances are 99 in 100 that it's an ISDN line, that is, one running the ISDN protocol. It's by no means everywhere yet, although its coverage is much higher in Europe than the United States, and higher still in Australia, Japan, and Singapore.

Although it was first designed in 1973, it's only recently that a significant number of U.S. metropolitan centers have had access to ISDN lines. To a large extent this slow adoption is because the phone companies must replace switches and install fiber cables. Another problem is that the standards have only in the last couple of years grown really solid, leading to the troublesome problem of having "flavors" of standards (which really means no standard at all).

You can think of ISDN as a somewhat slow LAN without the limits on distance and position that are inherent in LANs—a dial-up long-distance LAN, if you will.

What's ISDN Good For?

Until you have done a fair amount of remote computing, you won't have a good feeling for how much data you can transfer at 28.8 kbps. But consider one very popular application of remote access right now—exploring the Internet, especially its graphical multimedia layer, the World Wide Web (more about this in Chapters 9, 18, and 19). Multimedia involves a lot of data, orders of magnitude more than simpler text-based interfaces require. If you go to a Web site and select a musical clip to play, your Web software will first have to download that clip to your hard disk drive before it can play it. With a 2400 bps modem, this could take a full hour even for a relatively small music file. A 28.8 kbps modem is ten times faster. With a single ISDN line, you can double that speed—and so on.

In practice, most of the users of ISDN lines right now are businesses. They are placing video conferencing calls over the lines, running data-intensive connections between their various sites, and running much larger numbers of voice calls through single ISDN lines than they could through many more analog lines. There are some power users with ISDN lines in their homes, and the number is growing fast as the service becomes widely available, the price of the lines goes down, and the prices of connection equipment continue to drop. It will be quite commonplace before long. At present, though, it's still somewhat costly for the average user.

Bandwidth Increments

This section describes the common increments in which digital lines are available. The sizes are measured in bandwidth, but the terminology has lots of variations.

Basic Rate ISDN (BRI, BRISDN)

This is also sometimes called customer service ISDN. It represents entry-level digital connection capability. A single BRI line, known as a B (bearer) channel, can run at a data rate of 64 kbps. A full ISDN line consists of two B channels and a D channel. The latter runs at a rate of 16 kbps and is almost always used for signaling, which means your applications probably won't use it.

You can use just one of the B channels at a time to send your data, or use one B channel for data and the other for your voice traffic, or use both B channels at the same time for data, giving you a combined data rate of 128 kbps. This is how many of the higher quality ISDN-compatible video conferencing systems work.

Primary Rate ISDN (PRI, PRISDN)

When your communication applications get really greedy, such as when you install a multimedia WAN between two corporate branches, you need to start looking at options to get more than single ISDN lines at a time. PRI lets you choose much higher bandwidth increments of service. The North American PRI connection consists of 23 B channels and one 64 kbps D channel, resulting in a capacity of 1.536 megabits per second (Mbps). This size line is more commonly known as T1. In Europe, the comparable service offers 31 B channels, for a total speed of 2.048 Mbps (in the United States, this is commonly called European T1, or, more accurately, E1).

In some situations you may find that you need more capacity than a single BRI line offers, but you don't need a full T1 line yet. No problem; the network providers offer a service called *fractional T1,* which lets you specify some fraction of capacity between the two levels, in 64 kbps increments. It's possible to access this bandwidth on demand, by dialing up the network when you need it, or as a dedicated link that is always connected between your two sites.

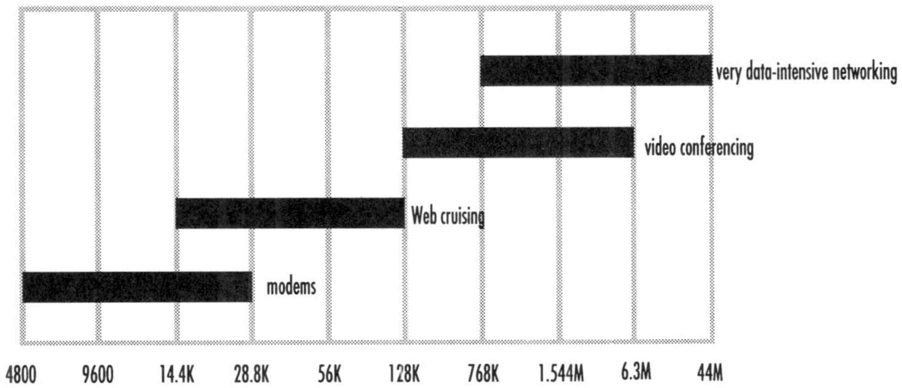

Figure 4.1 *Applications, data rates, and line types.*

In the other direction, if you need more than a single T1 line, providers offer T2, the equivalent of four T1 lines, giving 6.3 Mbps, and T3, which carries the equivalent of 28 T1 lines.

Figure 4.1 shows the range of data rates, the types of lines they require, and the most likely types of applications for the various ranges of capacity.

Setting Up for Digital Lines

The first step is finding out what digital options are available in your area. It's very likely that someone offers some means to get T3 service to you (for a price!), but if you need that level of connectivity, you probably already know who to call.

You're probably more interested in lower bandwidth options. To find out if ISDN is yet a reality in your area, call your local phone company's customer service division. They should at least know what you're talking about, and be able to give you a date that it might be available. If that doesn't work, try the commercial services division. In most areas, you pay for ISDN lines the same way as your current telephone service: an installation fee and a monthly service fee. These rates vary, but are typically several times the cost of ordinary voice telephone service.

Once you have established service, you need to buy equipment. Recall that your modem is designed to translate bits from the computer into waves that can travel over the analog network. On a digital network you don't need that; instead, you need a device that packages the bits from your computer into the packet structure used by the digital network. Such a device is called a *TA* (*terminal adapter*). As with modems, you can get internal or external varieties, that either come on a circuit board that goes inside your computer or in a box that sits near it.

There are other possibilities. If you plan to use your ISDN line for both voice and data, you need an ISDN telephone that also offers computer connections. Your service provider can advise you on the options, but a local electronics store can probably give you a better deal on the equipment.

Once your ISDN line is installed, you can either use both channels for a high-capacity data connection, or one for data and one for your voice traffic. You can connect your old analog telephones and fax machines to your ISDN line, as long as you have all the necessary interface hardware and software, but this is giving the lines short shrift. In addition to all the things your old analog line can do, your new ISDN line can, with the correct equipment, provide many features found in office PBXs, including multiple numbers per line, multiple lines per directory number, conferencing, forwarding, voice mail features, textual status displays, and much more, all because the digital lines are able to carry so much more information than their analog counterparts.

Digital Considerations

Once you have increased the size of the data "pipe" coming into your computer, you may still find that performance is not dramatically improved. The problem may not be with your ISDN connection or terminal adapter, but rather the port through which you are sending the data to your computer. Many serial ports struggle to keep up with 19.2 kbps, so you may end up wasting much of your new bandwidth unless you upgrade your computer hardware. If you install an internal TA, this will in many cases correct the problem, but check with the manufacturer.

Free from Wires

Traditional modems have another limitation that they share with telephones. They connect you not with a *person*, but with a *place*. When you call someone, you're generally not really calling the person, you're calling an office or a house instead.

This is changing, though, and wireless communications, especially cellular telephones, are leading the way. More and more, you can call someone no matter where she is (as long as she has her cellular phone with her and it's working). There were some 25 million cellular phones operating in the United States in 1995, up from 10 million in 1992. Futurists predict that in the next decade or two, you'll get a phone number that you keep all your life. Rather than changing your phone number when you change addresses, your phone number (or equivalent) will move with you. You'll use that number not only to speak to people, but also to send and receive messages, news, and video information, and shop. The vision is tantalizing, but the reality is currently somewhat different.

Portability at Last

Truly portable computers are becoming a reality. Four pounds (two kilograms) is still a lot of weight at the end of your arm, especially when you add all the accessories that they require, but it's a long way from the twenty-pound monsters of a few years ago. In a package that size, you can get a very high-powered processor, a color monitor of good resolution, and a keyboard that is 90 percent of the size of a standard desktop one.

Even smaller options include the so-called palmtop computers or PDAs (personal digital assistants). These do have to trim a few of the features of their larger cousins, but manufacturers are still managing to pack a lot into the small package.

What makes all of these entries so useful to remote access is that modems are also getting more portable. The homebound variety of modem resides on a board inside your desktop PC, or in a box beside it, making it rather awkward to move. The first answer was the pocket modem, that communicated with the computer through the serial port, was about the size of a pack of cigarettes, and was able to run from batteries if necessary, or from a small power supply. Right on its heels (and the current favorite) came even smaller modems, not much bigger than credit cards, that plug into the PCMCIA slots of those computers that have them. This variety takes its power directly from the PCMCIA slot, eliminating the need for batteries, power cables, and spare electrical outlets.

These solutions—notebook and palmtop computers, PDAs, super-portable modems—have all increased the mobility of the computer user dramatically. But they are just more convenient forms of the old modem solution, and thus still tie you to one place—wherever the phone line is. Wireless solutions are the answer. This section of the chapter first describes the current landscape of the wireless picture, and explains the problems lying in the way of the ideal situation. Next it describes what is available today as partial solutions to the problem of wireless data transport, in the way of cellular, radio, satellite, and infrared technology. Finally it focuses on a few specific products that are early ventures into the world of utterly portable computing, looking at some of the main entrants, describing typical features, and offering some tips on what to look for when buying one.

The Current Wireless Picture

Wireless transmission requires using the electromagnetic spectrum to transmit information. Figure 4.2 on page 39 shows the range of the electromagnetic spectrum, which includes audible, visible, and X-ray waves. Note that a large chunk of the spectrum, the radio spectrum, is usable for transmission with current technology. Users of the radio spectrum include television, emergency response teams, the military, cellular telephones, and garage door openers, among others. The higher the frequency, the shorter the range of the transmission, so the low end of the range is reserved for worldwide broadcasting. This is where the Voice of America radio station broadcasts, for instance.

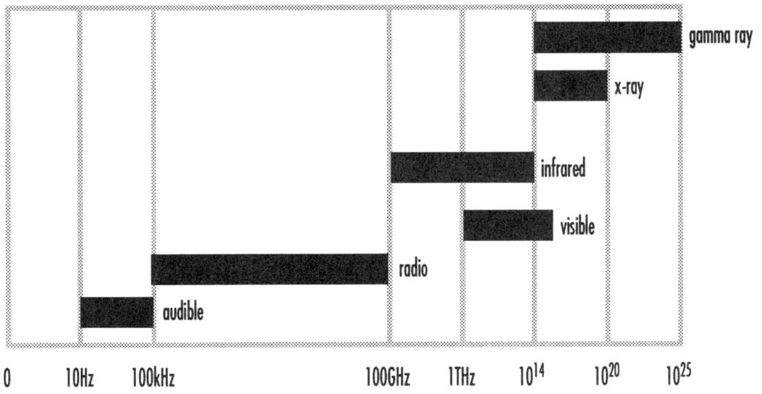

Figure 4.2 *The electromagnetic spectrum.*

Clearly, the radio spectrum is a limited resource, and world governments have the difficult job of splitting it up in a way that makes sense for consumers, technology, and existing equipment. In the United States, for instance, fire and police started out in the 30–50 MHz frequency band, but when that region filled up and more space was needed, the immediately adjacent area was already allocated to other applications. At that point emergency services were assigned part of the band from 150 to 174 MHz. This resulted in the problem of fire trucks from neighboring towns being unable to talk to each other. The solution at first glance would appear to be to reallocate the bandwidth so these services would have adjacent frequencies, but then everyone currently using that frequency would have to buy new equipment on both the broadcast and receiving ends, because the equipment only sends and receives at a single frequency. This would be very expensive and so seldom happens, but expect to hear more about frequency reallocation in coming years as providers and regulators wrestle with the problem.

This problem is parallel to the one facing truly global cellular transmission. In much of the world, cellular transmission runs over a network called GSM that operates in the 900 MHz frequency range. In the United States, though, this range was already taken when GSM was put in place, so U.S. cellular phones operate in a different range. This means that if you buy a cellular phone in the United States, you won't be able to use it in Europe (and vice versa). Some companies offer services that let you rent phones in Europe when you arrive, and forward all your calls to their providers, but this is not the hassle-free arrangement we hope for in the future.

Wireless Solutions Today

Data over Cellular Networks. The most straightforward way to achieve data transmission over the cellular network is to get a cellular modem and connect a cellular phone to one side of it and a computer to the other. This arrangement has the benefit of being relatively simple to set up, because the cellular modem mimics your old wire modem, meaning that you don't need any special software to connect to the network; you can still use your current communications software.

With this arrangement, you should be able to send and receive data from anywhere you can make a cellular call, just as if you were using your wire modem. There are some problems with the picture, though. The cellular environment is fairly hostile to data. As you travel behind hills or buildings or farther from the nearest tower, you may experience fading; when you move between cells there may be a brief drop in the call while the system "hops" to the next cell; and even connecting can be difficult because of the differences in how telephones and cell phones place calls. You may not notice most of this activity on the line during a voice call, but data calls and communications software are much less forgiving than the human ear. People using this type of connection for data report data speeds of 2400 on the average, 4800 in ideal conditions. They also report that successful connections are rather infrequent compared to the number of attempts.

Some providers offer modem pools to try to minimize the problems. This means you use your cell modem to dial the provider and send your data; from there they send the information to your destination over land lines, eliminating half of the cellular problems, and requiring you to buy a cell modem for only one end of the connection. The modem pool feature is also known as *DATA service because you must preface the phone number with that string in order to use your carrier's modem pool. This can be inconvenient if you want to use the same phone number list with both a land modem and your cellular assembly.

One way providers are attempting to improve the situation is by developing error-correcting protocols; two of the most common are MNP-10 and ETC. The former must be in both your modem and the one at the remote end; ETC prefers that arrangement but can operate adequately when talking to a non-ETC modem. Both improve the reliability of cellular data transmission, and are available in modems from a number of manufacturers.

Cellular Data in Packet Format. An alternative to trying to use the voice cellular network is to take advantage of research that is going into sending packetized data over cellular networks. A number of companies now offer data transmission that uses a digital packet technique called CDPD (cellular digital packet data), first developed at IBM. It solves many of the reliability and speed problems of data over voice cellular networks, but requires you to purchase special software and hardware, and requires the cellular provider to install compatible hardware as well. Once in place, it offers greater reliability than most cellular data mechanisms, but you can only communicate with other systems using the same technology. Thus you can't use it to send to modems or fax machines. If you always exchange data with the same place—the office LAN, for instance—it's something to consider. But if you need connectivity to a variety of telephone-reachable devices, this technology won't help.

Clearly, using the cellular network for modem connections is not as easy as using the much more mature land telephone system. Some of the problems are solved by using a more established technology: radio networks.

Radio Transmission. One problem of cellular networks is that there is no nationwide, much less worldwide, network. If you leave your network, you must either stop transmitting or pay a roaming fee for the privilege of using another network. This is one problem that radio networks don't have; they are seamless nationwide, and you pay a flat rate for access, with a per-packet fee for the amount of data transmitted. Another advantage is that they were originally developed explicitly for data transmission, so transmission is much more reliable and faster than is possible over cellular. For this reason, it is the technology of choice in most of the PDAs that offer built-in wireless data communications. In order to make this work, you need a special radio modem and compatible software. The two big U.S. networks are RAM Mobile Data and ARDIS. While the coverage of these is not as complete as the combined coverage of all of the cellular networks, the fact that they are nationwide tilts the scale. They are still not worldwide though; read on.

Satellite Transmission. For even wider coverage, you can purchase your own satellite dish and transmitter. Logistical problems abound with this solution, however. Whenever you want to receive messages, you must know your exact latitude and longitude, and compute the correct angle to aim the dish at the satellite (there are computer programs that will help you with this). On the sending side, you have to carry a large transmitter around with you. This is the solution of choice for journalists, relief workers, military personnel, or researchers far from cellular or radio networks, but the casual user may want to consider other options.

Two factors currently keep this technology from being more ubiquitous. One is cost. The satellites are extremely costly to launch; an entire satellite network costs billions of dollars. The other factor is portability. The equipment necessary for the transmitter to communicate with the distant satellite is currently heavy and power-hungry; most users don't want to carry around a generator, and the battery packs are also quite heavy. Despite recent advances that have shrunk the whole package down to about the size of a briefcase, they are still not very convenient for the casual user, and the devices cost thousands of dollars.

The next step is to get these devices down to a size that is comfortable to hold in the hand, more the current size of a cellular phone. This is possible now, but the resulting devices will be weaker than the current models. To answer this, a new network of lower satellites must be launched. This multibillion dollar project is the focus of several companies, including a consortium between Motorola, Loral Corp., and Inmarsat of London.

Whatever happens, satellite communications appear to have one fairly intractable problem: there is a built-in delay in the circuit that can run as high as a half a second. Computer protocols don't expect delays like this—they can malfunction if they don't compensate. Present efforts to compensate for this are not perfect, and even once they are perfected, they could result in slower transmissions than land lines. The other problem, with the current networks at least, is that the receiver stations are scarce. Until users start carrying receivers around with them, they have to make phone calls to the nearest stations, which can get expensive—not to mention the inconvenience of needing to know the phone number of the nearest station.

Infrared Transmission. You recall from the discussion of the electromagnetic spectrum that as a device operates on a higher frequency, the range at which it can operate shrinks. Infrared (IR) transmissions fall at the far right of the radio spectrum, and so have the shortest range of the wireless options. For this reason they are not practical for transmitting your reports to the office, but are used instead to eliminate the wires within the office. Their range is limited to about 850 feet (around 250 meters), and the two devices must be in a direct, unobstructed line from one another. This is the same technology your TV remote control uses. Infrared devices can transmit data at speeds up to 115 kbps, much faster than your V.34 modem's top speed of 28.8 kbps.

You can now find IR-equipped printers, mice, graphics tablets, and keyboards, all of which can go a long way toward reducing office wire clutter. Especially useful are IR devices in your notebook and desktop computers, combined with software that automatically updates the systems to each other (according to your specifications) when the two computers are in range of each other. Chapter 15 has the details.

When buying an IR-enabled device, make sure it incorporates the IrDA standard. Doing so should protect the future of your investment, since it seems to be fast becoming the industry standard.

Applying the Wireless Capabilities

Personal digital assistants, or PDAs, are considered the younger siblings of notebook computers, and until recently have been taken about as seriously as younger siblings often are. They promise to read your handwriting, track your appointments, finances, and phone calls, remind you of birthdays, and perform other functions of a combined calendar, address book, secretary, and accountant, all in a package about the size of your checkbook.

These devices have been widely touted as the next generation in computer accessibility, but the initial reception was rather lukewarm. One reason was the handwriting recognition feature. It appeared in these devices because they are aimed at people who do not use computers, or who don't like to use keyboards to operate them. (There's also a practical reason: keyboards do not readily fit on these devices.) The early implementations of handwriting recognition, though, have been somewhat disappointing. While some of them do "learn" and get better at recognizing your particular method of forming letters over time, the error rate can still be quite high, and many enthusiastic early adopters finally gave up. Most of the second generation offerings leave the handwriting recognition feature off, and instead offer a keypad for text input.

Certainly by going from a computer-oriented model of input in the keyboard to an input model that more closely resembles the devices it is designed to replace (that is, calendars, diaries, and checkbooks), all while keeping the price somewhere between $500 and $1,000, the industry has the right idea. There's still plenty of room for improvement, but the ground has been broken. This section describes some things you need to consider before buying a PDA.

Operating systems. The two most common are Apple's Newton operating system, used on the Newton MessagePad and the Motorola Marco, among others; and the MagicCap operating system from General Magic, used on the Motorola Envoy and Sony's MagicLink. Without a common operating system, two PDAs may not be able to exchange files with each other.

Infrared communications. There is no industry-wide standard yet, but many manufacturers use IrDA to send and receive infrared data. These include not only PDA manufacturers, but also makers of notebook computers and printers. The IrDA standard infrared transceivers appear on the IBM ThinkPad, the OmniBook from Hewlett-Packard, and the MessagePad from Apple, among others.

File exchanging. You can exchange files between your PDA and your PC, but there are some limitations. With many units you have to buy an extra package for the purpose containing the file exchange software and perhaps a serial cable (for those that don't transfer files via infrared signals). In some cases, file exchange is limited to simple ASCII text files—this means you may not be able to transfer your address book, for instance, or your financial data.

Modem communications. Some PDAs, including Motorola's Envoy, come with radio modems built in, while others, such as the Newton MessagePad, offer PCMCIA slots into which you can plug separate modems. If you plan to do a lot of large modem transfers, take a close look at the speed of the data part of the modem; some only offer 2400 bps. Also, look for one that uses radio transmission rather than cellular, for greater reliability.

Software. Graffiti, from Palm Computing, is considered to be the most accurate of the handwriting recognition software offerings. Typically, PDAs also come with database software, a scheduler, an e-mail application, finance software (usually Intuit's Pocket Quicken), and an editor.

Power. Small portable devices still need electricity from somewhere. How long are the batteries rated to last? Does it run on AA batteries, battery packs, or both? Nickel cadmium batteries are on the way out because of their high susceptibility to the "memory" problem: recharge them when they're only half empty and eventually they hold only a half charge. Nickel hydride batteries exhibit this problem too, but to a much lesser extent. Lithium ion batteries currently represent the leading edge, offering a much more lightweight, longer lasting charge, but these are not yet widely available.

SUMMARY

The applications of digital and wireless communications are many, and more become possible all the time. Since both arenas are still relatively new, though, the standards are not yet fully solid and choosing the right solution means weighing price, flexibility, availability, and compatibility. Wireless technology in particular is in something of disarray and changing fast, so if you plan to make an investment in it, you should strive for a two-year return on your investment. By that time, something quite different, much more useful, and cheaper is likely to be available.

chapter 5

SOFTWARE SURVEY

A WORLD OF OPTIONS

No matter what you want to do with your remote connection, there is software out there to help you accomplish your task. There are two factors you must take into account: the type of connection (digital, analog, over a cellular link or a modem link), and what you want to do over that connection. The last few chapters have introduced you to the issues in determining the type of connection. The second part, defining what you want to do, can be more difficult. This chapter surveys the possible applications, provides a guide to the limitations that the various connection types might impose, and offers some suggestions on features to look for when you choose your connection software. The information here will provide a basis for starting to understand what you need; later chapters that examine various specific applications will fill in the details.

WHAT CONNECTION SOFTWARE DOES

The details of what the connection software must do depend on the type of connection and the application to a large extent, but all implementations must perform some variation of the following steps: initialize the hardware, initiate the connection, manage the data exchange, and terminate the connection. They may also offer some features that help you manage your data outside the context of the connection. This section describes each of these functions.

Initialize the Hardware

In Chapter 3 there was a description of the sequence involved in a typical modem session. The first step in that sequence is to initialize the modem, telling it what data rate, compression, and error correction to use during the call, by sending the modem an initialization string. With other types of connection hardware this operation will be different, but they all have some version of it. Some connection software packages might perform the initialization when you first call up the software, while others wait until you start the call.

Initiate the Connection

Once you're ready to connect to the other device, the communications software initiates the connection. On a modem, this means taking the phone off hook, making sure there's a dial tone, dialing the number, and dealing with such difficulties as busy signals or non-answering remote modems. Once the two modems connect, they must exchange information regarding what data rate to use and what error correction and compression if any. This exchange process is called *handshaking*. When the two modems have agreed on how to converse, the handshaking phase is complete, and they report that fact back to the communications software, so it can take over with the data exchange. Many software packages let you do this all manually, entering the initialization strings and phone numbers by hand, but you'll nearly always want to arrange to run it all automatically.

The mechanism to place a call on cellular phones, though, is somewhat different, because there is no dial tone. When you place a cellular phone call, you enter the number and press SEND; after some clicks and pops, you hear the other phone ringing and from there things proceed as above. Since most modem software packages expect a dial tone before they begin automatic dialing, you may not be able to use them with a cellular modem. Happily, most cellular modems come with their own software.

Manage the Data Exchange

Once the two ends have established their connection, they can get down to the business of performing the requested tasks. This happens at two levels.

On the basic level, they just send data back and forth, make sure everything gets across the link without errors and without overwhelming either end, and acting as "bucket brigade" between the communications hardware and the applications level. You should never need to know much about this layer at all.

This second level, the applications layer, is the one you interact with. It takes the data from the lower level and organizes it into a form that is meaningful to you. This could mean assembling blocks into the file you requested, sending a fax, presenting menu choices from the remote computer and carrying your selections back, emulating a terminal that the remote computer recognizes, communicating with your workplace LAN, or presenting the client side of a client/server access system.

Terminate the Connection

This step is to inform the remote computer that the session is finished. You could just skip this step, unplugging the connection without explicitly terminating, but that's not a good idea for several reasons. First, unless you are connecting via a local telephone connection in the United States, you are paying for every minute (or 30 seconds) you stay on the line. If you don't explicitly hang up, the meter keeps running. This goes for long-distance calls, digital networks, and cellular calls. There's often also a meter running at the remote end, because many on-line services charge you by the amount of time you spend on-line.

The second reason to disconnect cleanly is that the remote computer needs to be able to close files, shut down connection processes, and generally tidy everything up. Leaving unattended connection sessions lying around is not only unfriendly, it can cause you problems later, such as not being able to reconnect to the port because the remote machine thinks a session is already running there.

Sometimes calls get accidentally disconnected. Most connection software is prepared for this, and has mechanisms for noticing that your session is no longer active and closing down everything related to the connection, but this process is not bulletproof and you shouldn't rely on it as a disconnection strategy.

Terminating cleanly also gives your communications software a chance to clean things up on your local computer, such as closing files and releasing ports. Not doing so can leave you with local messes, including corrupted file systems. If termination happens by accident (the power goes out, for instance), it's usually possible to recover, but it's best to avoid it whenever possible.

Manage Off-line Resources

Some communications software manages only the call itself. If you send or retrieve any data during the call, you must manage it outside of the context of the communications program.

More and more, though, the functions are integrated, so you don't have to know the distinction between the two layers of software. Packages are available, for instance, that let you compose e-mail messages, send them and retrieve them, organize them into folders, create address books and mailing lists, and archive messages for storage. The alternative, if you had separate communications and e-mail management programs, would be to retrieve each e-mail message as an individual file and organize it according to some structure appropriate to your operating system.

Another example is drag-and-drop style file transfer software. You connect to the remote system, click on the file you want to retrieve, and drag its icon to somewhere on your local system. The operations of getting the file over the link and placing it on your local system appear to you to be a single act. This level of ease requires careful integration on the part of the manufacturers of both systems, and is not available between every pair of systems you might want to connect.

Features to Look For

Now that you know what connection software does, you probably have some idea what sorts of features will help make it easy to use. When evaluating features, consider your application. If you expect to mostly send and receive e-mail, concentrate on features that simplify management of addresses and volumes of messages (Chapters 6 and 10 offer some suggestions). If you plan to spend most of your time exploring the World Wide Web, on the other hand, you'll want software that maximizes throughput while you're on line. The items listed below, though, are a good starting point.

Directory Maintenance

Since much of what you do with connection software is connect to other computers, it's very helpful to have a powerful way to maintain a list of places you frequently connect to. You need to be able to easily add, delete, and change entries, and call up an entry for another system and connect to it without going through a lot of steps. On the other hand, for systems that you only want to call once, you should be able to just enter that connection information without having to create a directory entry for it first.

File Exchange

This has two dimensions. First, you may want to be able to upload the address book in your PDA connection software, for instance, to the connection software you use on your PC. Second, if you already have an address book in some other format, it's very tiresome to have to type it all into the directory of the new software; you'll want an import/export feature instead.

Easy Ways to Find the Next Contact Point

If you are traveling and move between two satellite downlinks, you may need to know which to call to avoid paying astronomically higher long-distance charges. Or perhaps you are on vacation in Europe and want to know the nearest node. These questions are not only relevant to evaluating service providers, but also to summing up highly integrated packages of services, hardware, and software.

E-Mail Maintenance

Look for the ability to sort messages, search within the message header or body, create folders, automatically route or store specified types of messages, create mailing lists, and specify attachments. Also consider the mailing address book capabilities.

File Transfer Protocols

All major communications packages offer the most widely used transfer mechanisms. But if you know you'll be needing to exchange files using an exotic protocol, make sure the software offers it. More detail on these protocols appears in Chapters 8 and 14.

Ease of Use

This is a subjective call. If you can, experiment with the software a bit and see if its organization makes sense to you. Ask friends and colleagues about their software and how they like it. If you've never used any communications software before, many of the things won't be familiar, making it hard to evaluate. Ask for a demonstration and during it, ask to do the things you do know about. For modem software, ask how to change the initialization string. Enter connection information for a remote site. If you have to go through many levels or call up hidden screens, you may want to keep looking.

Software Considerations

With these guidelines, you're better equipped to select some communications software, but there are other issues that are specific to the type of connection you'll be making. This section looks at each of the most common connection types, and offers some suggestions on what to look for in the associated software.

Modem Software

Nearly all of the preceding discussions apply to standard modem software, including the caveat that what you select depends on what you want to do. The number of choices is enormous and growing, but here are the basic categories.

- *Generic connection software.* You can use simple connection software to dial up just about any modem in the world. These packages tend to be low on automation features but very high on flexibility. They offer little or no management of e-mail or other application-specific data, but usually have good connection information management. Example: ProComm Plus from DataStorm Technologies, Inc.

- *Software supplied by on-line service providers.* These packages usually automate the actions associated with a particular on-line service. They include capability to manage the messages and files common to that service, as well as information on connecting to the service itself. These features vary from limited to excellent. With some services, you have a selection of packages to choose from. Example: OzCis or NavCis for CompuServe. Most of the other on-line providers such as Prodigy, America Online, and Delphi have their own equivalents.
- *Software that comes with your operating environment.* These packages tend to be limited, but their big advantage is they cost you nothing more. Windows, for instance, is shipped on many new PCs, and it comes with its own connection software applet. It's not as fast or feature-rich as the third party solutions, or as automated as software from the service providers, but it might do in a pinch.

Within those categories, there are still dozens of features that you may or may not need. If your modem is a fax/modem, make sure the connection software you plan to use can send and receive faxes. If so, you might want to look for features that let you send several faxes at once to different numbers, delay them until the phone rates drop in the evening, retry if the line is busy, and do it all in the background while you go on to work on other things.

If you ever need to connect to a mainframe computer, you'll need software that provides terminal emulation. This is software that receives and sends the same commands with the mainframe that a simple terminal would, convincing the mainframe that it's talking to a terminal, rather than a PC with a brain of its own.

If you are planning to call lots of different types of systems with various types of on-line interaction methods, you need a generic software package. The one that comes with your modem or operating system may be adequate, but for heavy or complex use, look for one that can at least capture what's going by on the screen as it scrolls past, and that can let you write your own automation scripts if you think you might ever be so inclined.

Cellular Modem Software

The offerings in this field are sparse compared to the possibilities available for the more common variety of modem. In some cases, though, you may be able to use your cellular modem with software designed for ordinary modems. The potential difficulty, as described earlier, is the lack of dial tone in the cellular call, possibly forcing you to initiate the call by hand.

You probably won't have to, though. Most cell modems come with their own software; in fact, in the case of PDAs, the whole system is completely integrated: computer, cellular or radio modem, and software all come together.

Software That Comes with PDAs

Since PDA systems themselves are so small, the software that runs on them has to be small, too. At the same time, users expect (or manufacturers expect them to expect) software rich in features. While the software that comes on your PDA is probably not going to have all the features of its counterpart on your PC, the modem software guidelines above still offer a starting point.

In some PDAs, such as those based on the MagicCap operating system from General Magic, some of the communications software is part of the operating system. Sending wireless e-mail from these devices is a built-in capability.

Other functions might require you to buy an extra package. To send files (text only) from your Motorola Marco, for instance, to your Apple Macintosh, you must purchase a separate package containing file exchange software, and run the transfer over a serial cable.

Digital Connection Software

Whether you want to run a 56 kbps connection to the Internet over one of your BRISDN B channels, or operate a T1 link between the two main offices of your company, digital connections are a completely different story from their analog cousins in terms of software.

With a digital connection, you must first establish a link to the network. In the vast majority of cases, this will be a *TCP/IP* connection. TCP/IP is a network protocol that defines how the packets of data flow between the connected computers, and through them to other interconnected computers. It is a highly adaptive and flexible protocol, and is used by most of the systems on the Internet (in fact, its acronym stands for Transmission Control Protocol/Internet Protocol).

The lowest layer of software in the digital connection arena generally comes in two parts: a driver to talk to your terminal interface hardware, and what's called a TCP/IP protocol stack. This second part handles all of the aspects of maintaining the IP protocols over the physical link.

Above that, your application is the driving factor. Internet connections offer one set of possible applications, including remote computer log-ins, file transfers, e-mail, and multimedia possibilities; while video conferencing, another high-bandwidth digital application, requires quite different software (and extra hardware).

SUMMARY

Communications software depends on the type of connection you are using and the hardware you are using to make it. Nevertheless, all communications software has some features in common, and this chapter described some of those features. When selecting connection software, you have a huge array of choices; this chapter offers some suggestions on the most useful features to look for.

part II

Applying Remote Access

Putting Your Connections to Good Use

Part I of the book introduced you to remote access. It explored how the different possible technologies work, and described some of the many applications of remote access at a general level. This part of the book takes those descriptions further, examining some of the most popular applications of remote access in detail.

Chapter 6 starts off with a look at the "killer application" of remote access: e-mail. It's the most cited reason for installing LANs, is many people's main motivation for getting remote access, and is easily the most widely used of all remote applications.

A number of software and hardware strategies exist for connecting to another computer and running programs there, whether that remote computer is another PC or a mainframe. Chapter 7 describes these techniques and their differences, and offers some tips on selecting the option that's right for you.

Another frequent application of remote access is sending and receiving files between other computers and your own. Chapter 8 takes a look at what's involved, including dealing with the various types of formats and operating systems. It tells you what you need to know to make sure the files you get are useful to you.

Nearly everyone has at least heard of the Internet, but you may not be altogether sure what it's good for. Chapter 9 gives an overview of what it offers, the best applications of the different offerings, and some notes about its limitations.

This section of the book wraps up by coming full circle back to messaging, but this time around, messaging means much more than just e-mail. Chapter 10 covers more sophisticated message-based applications that represent the cutting edge of message technology.

This entire second part of the book describes applications in terms of how they work, what they're good for, when you might want to use them. Part III shifts the focus from what you can do to how to do it, so it focuses more on what you need to implement your remote connections and where to look to get software and help.

chapter 6

E-Mail Connections

Basic Messaging

Electronic mail, or e-mail, is arguably the most often used of all the applications of remote access. Next to printer sharing, it's one of the major reasons companies install LANs, and while it's probably not the main reason most people first buy a computer, it's certainly high on the list of reasons they begin to explore remote access.

If you haven't used e-mail much, you may be wondering what all the fuss is about. But those of you who have been using it for some time may find it indispensable. The ability to send messages to other people—whether your boss, your employees, your sister, or your old friends—not only frees you from the tyranny of having to be in a certain place at a certain time as for phone calls, but also lets you communicate much more information, faster, than is possible with some other messaging techniques such as pagers or voice mail.

One of the key advantages of e-mail is that it's *asynchronous.* This means that the sending and receiving parts of the e-mail process happen independently of each other. Your friend sends you an e-mail message in the afternoon reminding you of a lunch date. You are at the office in the afternoon, but find the message waiting for you when you check your messages at home that evening. Rather than being interrupted by a phone call, you can process the request when it's convenient for you.

This can also be one of the disadvantages of e-mail. You might send e-mail to someone, but not know whether they acted on your request or information or even whether they got the message. Thus it's not a complete replacement of synchronous communications mechanisms like telephoning, but a useful adjunct.

E-Mail Background

The first implementations of e-mail were on large multiuser computer systems. All of the users on a PDP-11 computer running UNIX could send messages to each other—not terribly useful, since they could also just walk across the room or down the hall and talk to each other. E-mail got more useful with the advent of networked computers. If you worked at IBM, for instance, you could use the in-house e-mail system to send a message to any other employee at IBM, whether in Armonk, New York, or Austin, Texas. The application you used to send the message, in both cases, was determined for you by the system administrators; you didn't get a choice. Neither did they, in the early days; the messaging feature of these systems was not an area that received a lot of attention.

When LANs started cropping up all over the landscape, e-mail began to be more important and more often used. LANs often were first installed to facilitate the sharing of expensive or seldom-used resources such as printers and disk space, but as more and more people sent messages over the LAN e-mail system, e-mail itself became an increasingly important part of the equation. In response, vendors started paying more attention to e-mail packages, and administrators and users began to have more options in terms of features and capabilities.

At the same time, users of on-line services were exchanging e-mail messages with each other. A CompuServe user, for instance, could send a message to any other CompuServe user in the world. While this was quite a lot of people, it was no help if he wanted to send e-mail to someone who was using the Internet, for instance, for her e-mail services.

This is where gateways come in. They transfer e-mail between systems, translating address formats where necessary. All of the major on-line services, and most of the smaller ones, have gateways that enable you to exchange e-mail with users of all the other major systems. Thus, someone who uses SprintNet for e-mail can send messages to Internet users, CompuServe users, America Online users, and so on.

With this interconnectivity, you can now exchange e-mail with millions of people. The explosive growth also means that you have lots of choices when selecting e-mail handling software.

How E-Mail Works

E-mail works via a *store and forward* mechanism. You send a message from your computer to an e-mail host, and it examines the address to decide where to send the message. If it is addressed to a user on the same system, the host stores it until the addressee next checks his or her e-mail. If it needs to go to another system, the host forwards it to that system, or to the next one along the way to that one, as shown in Figure 6.1.

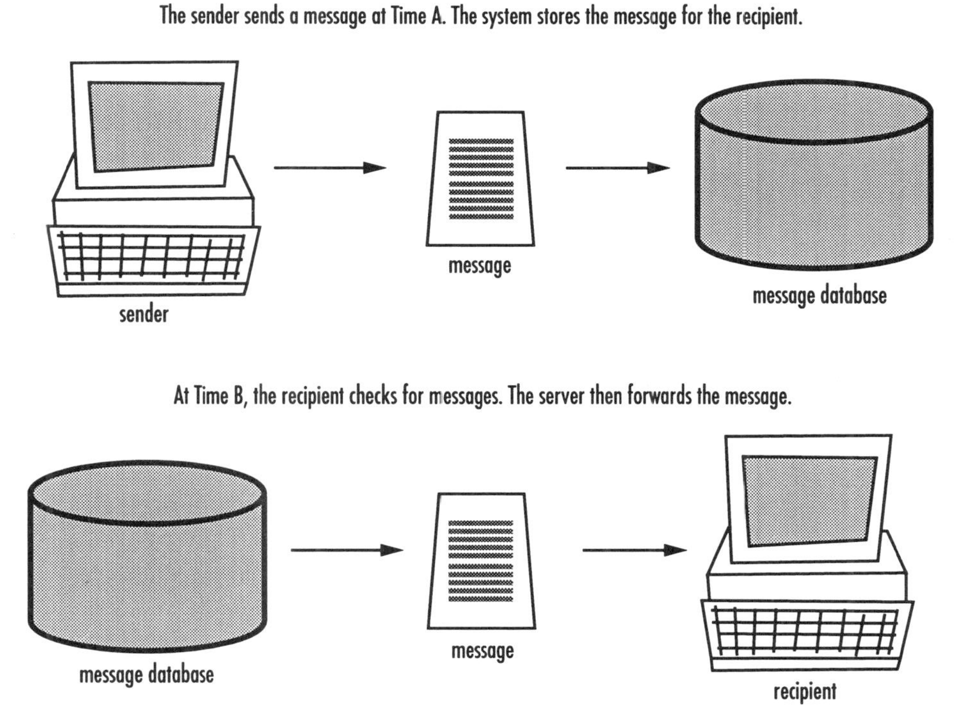

Figure 6.1 *E-mail messages are sent via a store and forward mechanism.*

When the message arrives at the addressee's host, what happens depends on the configuration. If he or she has arranged to be notified of incoming messages, a small message box may appear onscreen announcing the incoming e-mail, or the addressee may hear a beep or see a flashing icon. If he or she doesn't want to be bothered with notification (which can be not only disruptive, but in some implementations also wastes valuable system resources), or if he or she isn't always connected to the host computer, the addressee doesn't find out about the incoming message until he or she next connects to that computer to see whether any messages have arrived.

What happens to the message between the time it arrives at the host computer and the time the receiver retrieves it? From the time they get sent and during the time that the messages remain stored on the host, most implementations of e-mail systems make sure they are encrypted, so that the contents are not revealed by casual browsing. When the addressee retrieves messages, the host system unencrypts them and either displays them on his or her terminal or via the client process running on the desktop PC.

Features of E-Mail and E-Mail Software

Before you can start shopping for e-mail software, it's helpful to know what types of features are typically offered and what they're good for. This section describes the most common and most useful features.

Address Books

Since sending e-mail involves knowing the electronic address of the person to whom you're sending it, you'll need some method for organizing the addresses so you can find them quickly, add, change, and delete them easily, and sort or search on various fields. It can be helpful to be able to add a note or comment to each address record.

Message Forwarding

If you get a message and you know the contents would interest another of your correspondents, it's nice to be able to easily send them a copy of the message. This should not require you to save the message to a file, then create a new message using that file; it should be automated within the e-mail software.

Quoting

Sometimes when replying to a message it's helpful to include a quote from the original message to provide some context for your response. More and more e-mail systems provide a way to do this automatically, saving you from having to type the whole quote in again or performing laborious cut and paste operations via the operating system.

Mailing Lists

You often have the occasion to send the same message to a group of people. This might be all the friends you wish to invite to a party or all of your employees working on a particular project. Look for software that allows you to create and store such lists as subsets of your entire address book.

Public vs. Private Mailing Lists

Some systems allow you to create private mailing lists that only you use, and also public ones that are available to everyone. The latter are used mostly in work environments, so that everyone in a particular department, for instance, can easily send messages to everyone else in the department.

Folders

With all these copies of messages and broadcast messages going around, you're going to need a convenient way to store them. At the very least, the system should provide some means to store messages in groups, letting you create the groups (usually called folders). You might organize them by sender, project, date, or some other criterion—anything that will make them easy to find later.

Maintenance Features

You won't want to store all messages forever, though. You can get software that automatically culls everything after a specified length of time, or that archives messages in a compressed form. Look over archive capabilities carefully—many don't give you any means to delete messages from the archive or to view them without uncompressing the entire archive, so once they're stored in archives they become much harder to work with.

Sorting by Rules

Smarter e-mail systems let you define rules for the handling of certain messages. Perhaps everyone in your workplace, for instance, is in the habit of sending a message to the entire company when they will be out of the office for any length of time. Messages of this type can quickly become a blizzard, but may not interest you. If your software allows it, you can instruct it to automatically discard all messages with the phrase "out of the office" in them, and you never have to see them. An especially nice feature in this area is the ability to exclude messages using a rule; for instance, to specify that all messages that do not refer to a particular project be routed to a colleague.

Attachments

It's almost essential that the e-mail system be able to handle more than just text messages that you type into the body of the message. You should not only be able to attach text files to the message, but also binary files such as pictures, program files, or sound files.

Multimedia Formats

Perhaps you can attach pictures or sound files, but the receiver has to extract them and play them outside of the e-mail software. Sophisticated e-mail systems allow him or her to play the sound or display the graphics upon receipt. A more pedestrian feature but still handy is being able to send faxes from your e-mail system.

Encryption

Encryption is especially useful for sensitive messages, but so far this feature is rare in commercial LAN e-mail packages. If security is a concern, see also Chapter 24.

Import/Export Capability

If you already have an address book somewhere containing e-mail addresses that you can use on the new system, you'll probably want to be able to import them into the address book of the new system. Typing them all in again by hand is no fun.

Your Feature Set

Whether you have any or all of these features depends on three things: your client e-mail software that runs on your local machine, the server or host e-mail system on the computer you're connected to, and the standard to which your system adheres. Not all e-mail systems support all of these features.

Even if your system does support every one of these features, you may not be able to exchange all of these items with users of other systems. This depends on both the messaging standard used on the other system, and the capabilities of your gateway between the two systems. The next section explains.

Talking to Each Other

E-mail presents the same problem that comes up with any other type of communications—being able to communicate with those who speak a different language. Many e-mail systems evolved in relative isolation, and thus developed their own particular ways of representing addresses and specifying delivery. At first glance, it might seem that you have to have an address on every possible e-mail system in the world in order to communicate with users on every system. Though some people do maintain e-mail accounts on many different systems, this is not necessary because standards have arrived, and when two systems running different standards (yes, so far there is more than one standard), the systems can communicate with each other through a device called a gateway.

E-Mail Standards

There are dozens of different e-mail addressing schemes, but only two are widespread enough to be called standards. The one most widely used in the United States and Canada is formatted according to a standard called RFC-822. Since this is the mechanism used by e-mail systems on the Internet, it's also called the Internet standard. Though it has some weaknesses and limitations, it has on its side a large body of software that already uses it, all over the world. For that reason, it's favored to become the worldwide standard.

The other big contender has the law on its side, or more specifically the ITU-T. This is the X.400 standard. It's not strictly correct to call it MHS (message handling service) but many people do, because the MHS protocol is based on the X.400 addressing scheme. This addressing standard is the one Novell has adopted, so it's used in a great many LAN e-mail systems all over the world. It's also the one favored by the International Standards Organization (ISO) and appears in many European e-mail systems. Its big weakness against the Internet standard is that there is relatively little software actually using it compared to the large body using the Internet addressing scheme.

Gateways and Mail Switches

In order to send a message out from your Bitnet account, for instance, to someone on an Internet account, the message has to undergo some sort of translation. This translation is performed by a gateway machine, which receives the message in the local format, converts it to the remote format, and sends it on its way.

If you're interacting with the e-mail system only at the user level, that's all you need to know. If you're the network administrator in charge of implementing connections to outside e-mail systems, though, deciding how to handle gateways can be more complex. If you only need to connect to one other type of e-mail system—from a Novell LAN to Internet e-mail, for instance—a single gateway is sufficient. But if you also need to handle other types of messages, gateways quickly become cumbersome—you need a different dedicated PC to act as a gateway for every message type. Use a *mail switch* instead; these are also known as multiple gateways, because they can provide translation between more than just two types of messaging schemes.

A significant limitation of most gateways is that they do not allow for attachments to messages, and do not provide a mechanism for sending nontext messages. The MIME standard (multipurpose Internet mail extension) attempts to standardize the business of attachments; under this scheme, if a message arrives with an attachment, the mail client determines what type of file the attachment is and starts up the application that you need to view it. Note that this presents some dramatic security holes, as outlined in Chapter 24.

E-Mail Software Options

The software you use to read, send, and store e-mail depends partly on your e-mail provider. This section discusses the typical categories, and what you can do about setting yourself up with a nice e-mail system.

LAN E-Mail Software

If your primary e-mail source is the office LAN, you are probably going to have to settle for whatever e-mail client your network administrator has selected. This is actually not such a hardship, as there are many excellent packages available. Some typical options, in no particular order, include:

- cc:Mail from Lotus,
- GroupWise from Novell, and
- Microsoft Mail from Microsoft.

If you're the network administrator, though, and are evaluating new packages, consider the features listed in the section "Features of E-Mail and E-Mail Software" on page 60 before you buy.

Free Access E-Mail Software

This is a huge category; it includes all of the public on-line services and the Internet. The on-line services alone can each offer a dozen different alternatives for accessing their systems; most of these include e-mail management. CompuServe users, for instance, can choose from Macintosh-based, Windows-based, or DOS-based versions of software that automates a CompuServe session to varying degrees. One of the tasks they all perform is e-mail handling: retrieving, answering, creating, sending, and storing messages. You'll probably base your selection not only on the e-mail features but on the level of automation, the platform, and the quality of the user interface, in these cases. Other on-line services do not offer so many different options of software interfaces; the technical support people on the particular forum will be able to enlighten you on what is available.

On the Internet side of things, the field of choice is even wider. The options available to you depend on whether you are able to run a graphical interface on your Internet account, or must rely on the text based version, at the most basic. If you hold an account on an Internet system already, ask the administrators there what the local options are. If you are administering such a site, you should be aware that some of the more popular and powerful options are riddled with security holes; these are discussed in Chapter 24.

Wireless Messaging Solutions

If you only plan to receive messages, but you need the portability of a wireless solution, consider a pager-style connection. One such is the PageCard Wireless Messaging System from Socket Communications. For a couple hundred dollars, you get a receiver that plugs into a PCMCIA slot, a card that displays the brief messages and that can store longer ones. You pay a monthly fee for the service.

Another wireless messaging service is RadioMail. Like Socket Communications, it provides one-way messaging, but it also offers a two-way send and receive e-mail service. You need a radio modem to use the service.

Communicating via E-Mail

At first glance, e-mail seems not much different from paper mail. But it is so easy to fire off a message or a witty response or a tidbit of gossip via e-mail; so many things get said over e-mail that would never get said on paper.

E-mail is not paper communication, and it is not face-to-face communication. The clever retort or observation that would be placed firmly in context face to face with an ironic raise of the eyebrow can sound very hurtful over e-mail. Further, since e-mail is so easy to reproduce and forward, you might find that your amusing comment about the boss's outfit gets very quickly to the very person you would least want to see it.

Communicating effectively via e-mail does mean learning a new communication style. A little bit of forethought can go a long way toward preventing mistakes.

Summary

E-mail gives you a tremendous amount of flexibility in staying connected to your colleagues, family, or friends, no matter where you are, or when you decide to check in. But it only lays the groundwork for more sophisticated messaging applications, such as workflow automation and groupware. Chapter 10 covers this fast-growing aspect of messaging systems in more detail.

chapter 7

Two Places at Once

Using Your Computer to Work Elsewhere

One interesting option in the world of remote connections is the capability that some software packages provide to let you use your computer to control another. This other computer could be across the room, on the other side of the building, or in another city. You could connect to it through a direct connection between the two computers, over the LAN in the building, or over the phone lines. The degrees of control you have over the other computer vary—from only being able to view files in a prearranged directory, to being able to completely control the keyboard, mouse, and screen of the remote computer.

This chapter surveys software that provides these types of capabilities. The ways in which you can connect to remote computers vary as widely as the things you can do, so the chapter first presents a general overview of remote connection capability. Then we'll examine the different types of connections in terms of the types of computers and connections involved, the degrees of control you have over the remote computer, and the nature of the interaction.

Types of Remote Connections

One of the most important distinctions between the different types of remote connections is the amount of control you have over the remote computer. In some types of connections your local computer acts as a simple terminal and the remote computer does all the work. You use your computer to initiate commands and processes on the remote computer, and it reports back to you the results of those processes.

In other types of connections, you have a bit more control. You might, for instance, be able to view the files on the remote computer, but perhaps only those in a certain part of the file system. In this type of connection, you typically have a limited set of commands that you are allowed to use on the remote computer. These usually include commands to move around in the file system, copy files to your local system, and send files to the remote system.

Moving up a bit on the scale gives you complete control of the other computer. In this type of connection, you control the mouse and keyboard of the remote system, and the on-screen results of your input there are displayed on the screen of your local system. In some such products, you can display the results of your remote session in a subwindow on your local screen and you can still access all of the resources and applications on the local computer; in others, the entire screen of your local computer is dedicated to displaying the remote session, and access to your local system may be restricted during the course of the remote connection.

Another type of remote connection is the fully interactive variety. Rather than attaching to a remote system that is probably unattended, and using it as an extension of your local workspace, this type of connection focuses on the ability to attach to another person's system so that the two of you (or more) can work together on the same application or document. Usually, in order to be useful, the parties working together would also be conversing over a telephone, discussing the changes they were making via computer as they made them.

The rest of the chapter examines the most commonly used forms of these types of connections.

PC-to-PC Control

There is a huge array of software available that allows you to connect two computers. The capabilities range from just transmitting files to taking full control of the remote computer. Here's an overview of these products; see Chapter 15 for a full rundown.

Remote File Transfer

This type of software is most useful for sharing a limited number of files with other users. Typically you specify beforehand what files the remote user will be allowed to see, or what subtree of the file system, then run the software that implements the connection. The other user connects to your computer and can copy the authorized files to her own computer, or, if allowed, can copy files from her computer to yours. This is usually much faster and more efficient than copying files to floppy disks and then to the other computer, especially if the quantity of data to transfer fills more than a single diskette.

Another common use for this capability is to synchronize the data between your laptop or notebook computer and your desktop computer. Keeping two sets of files, one on each computer, can result in a high number of merging headaches—products such as LapLink from Traveling Software have evolved to minimize the attendant hassles.

Remote Connection and Control Software

These vary in capabilities and complexity, but they all generally offer the ability to connect to another similarly configured PC and execute programs on it. Basic communications software products such as ProComm Plus from Datastorm Technologies let you configure the remote PC as a host; it then waits passively until you dial in to initiate the remote connection. From there you can enter commands on the remote computer, including performing file transfers.

Other products, such as Norton pcANYWHERE from Symantec, offer more elaborate control of the remote computer. With this type of product you have much fuller control over the remote computer, and can run programs that require graphical environments such as Windows.

These products are useful not only for remote work, but also in technical support situations, where trying to describe a problem or configuration over the telephone can be frustrating for caller and support technician alike. Companies that use remote control software for technical support report a significant decrease in support costs and happier customers, and many PC manufacturers are preloading remote control software on their systems for that reason.

Another common use for remote control software is in remote training. The trainer can work with classes at widely distributed sites, describing techniques and procedures while the class members follow along. This not only reduces travel time for trainers; it ensures greater uniformity between training sessions, and many corporations are adopting remote computing methods in their in-house training programs.

Remote Connections to a LAN

Sometimes you need to connect to a LAN when you are not in an area served by that LAN. There are a number of ways to accomplish this. You can add nodes to your LAN that allow limited access to it, either by adding remote dial-up nodes to the LAN server itself (one such product is WanderLink from Funk Software), or to a dedicated server that handles all the remote calls. The latter method may offer more security, because you can configure the dedicated server as a firewall (see Chapter 24 for more information about security and firewalls). Products that offer this capability include Remote LAN Node from DCA, and Remote Office Gold from Stampede.

Another possibility is to use remote control software, described in the previous section, to connect to one of the PCs on the LAN. You would configure this by attaching a modem to the particular PC itself, so that the remote user can dial into that PC only.

Finally, you can attach a modem to the server and use features of the network operating system to handle the remote connections. Novell NetWare, for instance, offers virtual machines to remote callers, allowing them to access all of the network resources as if they were logged in locally. Chapter 16 has more information about these options.

Remote Connections to a Mainframe

Not too long ago, you used a terminal to connect to a large mainframe computer, and that was about the size of computing. These terminals were called "dumb terminals" because they did no processing of the information coming from the mainframe computer, or of the input going into it; they just displayed your input and the mainframe's responses to it. Now though, many of these old mainframes have been replaced by networked groups of PCs, and even in environments where the mainframe remains as a central repository of the company's data, many people also have PCs on their desks.

For a while, people working in these environments had both a PC and a terminal, but this is now unnecessary clutter. Instead, you can drop a card into your PC to handle the connection to a mainframe, attach the old terminal cable to the card, and run some software on your PC to make it look for all practical purposes like a dumb terminal, from the point of view of the mainframe. In this configuration your PC acts like a terminal; it's said to *emulate* a terminal, and the software that performs this magic is called *terminal emulation software*. Chapter 17 goes into these types of connections in more detail.

Remote Conferencing

Remote conferencing products turn the usual one-way type of remote connection into a two way street. With them, rather than connecting to your office PC, or some other unattended remote computer, you link to another person's PC, and the two of you can then work together to share documents, both making changes to the documents at the same time. Usually these products work best when the two remotely linked workers are also talking to each other, so people using a remote computer connection to jointly edit a document typically augment the remote connection with a phone connection, and discuss the changes with each other as they work.

The capabilities of conferencing products vary. Some only allow you to take snapshots of screens before you make the connection and send those snapshots over the link during the session so that you can both discuss the same picture. Usually these types of products at least offer shared cursors, so you can both point to different features of the picture as you discuss it. More advanced features include shared drawing capabilities, so you can not only both point to areas of the pictures, but circle features, write on the picture, and develop diagrams together.

Up the scale of sophistication in conferencing products are features that let more than two people participate in a conference, that allow users to share not only graphics, but entire applications. Some allow you to link a document to another, using technology such as OLE (object linking and embedding) over the remote connection. With this feature, both participants can, for instance, draw up a document using a shared document editor, then one user can embed a spreadsheet table in the document so that both sites can see changes to the table for the duration of the link. An example of a conferencing system with these capabilities is Intel's ProShare.

These products are up against a lot of technical challenges, and can therefore be tricky to configure and a bit fragile. They work best when both systems are running the same set of colors over a fast link. Mismatched color palettes on the two computers can slow down the link considerably, and a slow link can mean that you spend more time during the session waiting for the conferencing software to update the screen than you do actually getting work done. Still, these products are improving, and within certain bounds can greatly enhance remote work.

You may wonder what the difference is between conferencing software and remote control software. The line is indeed blurry. Remote control software that allows the remote user to control her own computer, rather than completely taking over control of the remote computer, offers the most basic form of conferencing software. Most remote control products, though, don't allow the remote link to run in a subwindow on both systems while allowing users full access to their local systems. Essentially, in remote control software there is a clear distinction between the host and the client. In conferencing software that distinction remains, but the relation between the two computers is more on a peer level.

Security and Working Remotely

Setting up a modem on the office LAN opens up the LAN to more people than just you and those you authorize to use it. Once you can access a computer remotely, the potential exists that other people who have similar configurations to yours can do the same.

Whether you regard this as a problem, or an issue worth worrying about, depends on several things: what you have to lose on the remote computer, how often the remote computer must be available, and how much you are willing to invest in keeping intruders out.

This section serves as a brief discussion of the security issues that you face when you make a computer remotely available. You'll find a wider discussion in Chapter 24.

Considering Security Options

When you attach a modem to a computer or a LAN to enable you to reach that computer whenever you choose to, you also enable other people to access the computer as well. You might think that by not making the phone number public, you are restricting access to only yourself and those you tell the number to. If this is the case, probably the only other people who ever reach the remote connection modem are those who misdialed, the reasoning goes. While this makes sense, the reality is that there are some people who make a hobby of finding remotely accessible computers, with software that dials all the possible phone numbers in an exchange and makes a note of which numbers are answered by computers rather than people or fax machines. Then the individuals can go back later and explore all the computers that are available for remote connections.

Should this concern you? To some extent, the answer is "it depends." Because the risk of malicious activity is always present, the answer is yes, you need to protect your remotely available computers from unauthorized access. No matter how little value you may place on the material on that computer, losing it, or having the machine sabotaged, is a loss you'd rather avoid—otherwise, why place the computer on line at all? So while you might not be worried about losing sensitive corporate data because you don't store it on that machine, you still need to protect the machine from remote attacks. This doesn't happen often, compared to the number of systems out there, but it does happen. It makes no sense to lock the barn door after the horse is gone, as the saying goes.

Security Solutions

The simplest security strategy is to simply not leave the machine available all the time. If you know you will be dialing into the office LAN to access your office PC on a given afternoon, perhaps you can arrange with the network administrator to turn on the modem and activate the remote connection software just before you call, and disable it after you hang up. Similarly, if you only use your modem to make calls from your home PC, and you don't run software on that PC that allows other computers to call yours, you have nothing to worry about—no one can get into the machine in that case.

But this is not always practical. The point of having remote access is often being able to reach the remote computer at any particular time, without having to have someone at the other end managing the remote computer. All of the major remote connection software products offer features that help make this less risky.

The most common of these strategies, and one that can be fairly powerful if not misused, is the lowly password. A well-chosen password, changed often, can offer a fairly high barrier against even determined intruders. The key is choosing passwords well and changing them often, two things that most people don't bother with. See Chapter 24 for more detail.

Passwords offer a form of user identification. They are not bulletproof, since unauthorized persons can guess passwords. They do, however, increase the likelihood that the person logged in is an authorized user. The next level of security implemented by many of these products is to configure the remote system so that users accessing the system remotely can only reach certain directories or only see certain prespecified files. This works well in situations such as publishing files on line, where you know that all remote users only want to copy one or two particular files from your computer and then disconnect. If they try to access anything else on your system, the attempt fails.

Another ingenious technique is the *callback*. Systems that use this mechanism require that you specify beforehand the phone number of the system from which you will be calling. Thus you configure your office PC to accept remote connections before you leave for home in the evening, and tell the system the phone number to use to call your home PC. When you arrive home and are ready to get to work, you call the office computer. It answers, hangs up, and then dials your home number to make the connection. If anyone else calls your office PC, they initiate a call only to your number, which prevents them from accessing your system. Clearly this works best for situations when you always access the remote computer from the same place, but with some help from someone at the remote end you could also make it work on the road. It has the additional benefit of placing the costs of the telephone calls on the remote system's site, which might or might not be an advantage.

Another feature of some remote control software packages, including Norton pcANYWHERE, is the ability to configure the host side of the connection to blank the screen once the connection is established. While this would be undesirable in shared work situations, it might be quite handy if you're dialing into a computer in a relatively public place and don't want passersby viewing the course of your on-line session.

SUMMARY

This chapter briefly surveyed the possibilities available to you when you need to connect to some other computer in another location. Clearly, the type of connection you need and can use depends on several things: the type of remote computer, the type of connection available, the access you have to that computer, and what you want to accomplish. You now have an overview of what the choices are and how these restrictions affect your choices. Later chapters, especially in Part III, describe in more detail how to achieve these types of connections.

chapter 8

Getting Software and Other Files

Well-Hidden Treasure

You've probably heard so much about all of the wondrous offerings available on-line that it might start to seem like hype. If you have ever tried to wander out into the on-line repositories to look for something on a particular topic, you might feel sure it's hype. While it's certainly true that an incredible assortment and quantity of different types of software and information are available on-line, much of it publicly, getting it is usually not as simple as dialing a phone number with your modem and typing in a single "get" command. Compatibility differences abound, even after you have mastered the nontrivial task of actually finding what you want.

This chapter cuts through the confusion and helps you understand what's available. It first presents an overview of the types of files you can find, then describes what you need to know about the files before you can make use of them. Finally, it describes some risks and concerns that you need to watch out for when gathering files from public sources. Chapter 14 takes this discussion to a more concrete level, presenting some examples of search techniques and mechanisms, and describing the tools you need to hunt for files.

What's Available

This section describes in general terms what types of files you can expect to find when you go searching on-line for data and programs. It's pretty safe to say that no matter what you're looking for, either it is out there or you can find something similar available. Perhaps you already have a good idea what you want to look for, but whether you do or not, this section may tempt you with things you never knew you wanted.

Software

Name just about anything you might want to do with a software program, and chances are there is a version available somewhere on-line. From e-mail systems and operating systems to spreadsheets and war games, the array of software available from on-line services and sources has something for everyone.

How do you pay for software distributed this way? Some of this software is in the public domain, which means it is available to the public and you are not charged for using it. Many excellent programs are offered as shareware. Under this system, you download a program that interests you and try it out for a few weeks or a month. If in that time you find it to be useful, you are asked to remit a fee to the developer—an address is usually listed somewhere in the files that come with the software. The fees for this type of software are often relatively low, and payment is on the honor system. In return for paying, you get technical support, printed manuals, and/or discounts on upgrades. A number of popular software packages are distributed this way; some commercial software products, notably ProComm Plus, started out as shareware and were so successful that they moved to the shelves of software stores.

Some software is distributed via modem as a convenience for the customer. You call the company, give them a credit card number or other payment information, and they tell you where to download the software and perhaps give you an access code. This is not shareware; you have to pay for it before you get it, and for that you get the support and other benefits you expect when you pay for a shrink-wrapped product. While not very many companies distribute full packages this way, a great many offer patches and upgrades through a system like this.

Another type of software available on-line is computer source code. If you are a programmer, or want to become one, these files give you a chance to download code and compile it on your own system, perhaps customizing it with your own features and enhancements. In concept at least, this means you can get source code for a program that runs only on Macintoshes, for instance, and change it on your system to run under Windows. In practice this is a nontrivial task, and people more often port on-line source code between different versions of UNIX systems than between completely different operating environments.

Multimedia

One of the fastest growing categories of remote files is the collection of sound, pictures, and movie files on the various servers of the world. One of these categories, comprising picture files, has been around for a long time; the other two are more recent, because PCs and their peripheral equipment have only recently become fast enough to handle decoding and playing sound or motion picture clips.

This type of data is best accessed with an automated client such as an Internet World Wide Web (WWW) browser. These browsers (and the servers with which they interface) handle all of the details of determining whether a file is a sound, motion picture, still graphic, or text file, and displaying it or playing it accordingly. But when you want to get one of these files manually, such as when it's not attached to the Web, you need to know those details yourself; the section on compatibility later in this chapter fills you in.

Research Sources

A few years ago someone had the idea that it would be great if someone would put the text of the great works of Western literature into a database, so that researchers, scholars, and other interested parties could search the text for certain words, view it on a screen, and perform analysis on the contents. Believe it or not, this is happening, in a massive undertaking called the Gutenburg project.

Not only can you find works of literature on-line, but there are databases containing such things as all of the magazine articles published in the United States in the last several years, searchable by keyword. You'll find financial databases, where you can look for a company's annual report and any references to that company in the press over a given period of time. It's even possible to view graphical images of famous paintings.

Most of these types of resources are not free, though. To cover the cost of setting them up and maintaining them, you may be charged certain fees for the different services. Some of them charge a one-time fee to give you access, others charge by the month, but most charge by the hour (often in addition to the setup or monthly fees). Chapter 14 has more detail.

Miscellany

It's impossible to list all of the different types of things available via remote file transfers. From travelogues to catalogs, government reports, joke lists, travel information, and instructions on how to do just about anything from fly fishing to building compilers, it's out there. One trick is finding it; Chapter 14 covers that. Another trick is knowing what to do with the file once you've got it, and knowing whether you can do anything with it. That's the topic of the next section.

File Types and Compatibility Issues

File collection and file transfer software is getting more and more automated, and a lot of the matters discussed in this section may be handled for you in many cases. One can hope that someday soon this information will all be as unnecessary as it is now unnecessary to know how to key in the booting instruction codes on the front panel of the computer. Web browsers and servers, for instance, work together to shield you from all the details of file types, formats, and hardware requirements. You go to a Web page and click on a button to play a music clip. The browser transfers the file to your computer, decompresses it, and sends it off to your sound card to play it, all transparent to you.

For all this to work so smoothly, of course, the software has to know quite a bit about the files involved. The difficulties come up when you want to access data that is not organized into this type of format. When this is the case, you're the one who has to know all about the formats, compression types, display techniques, and transfer rules. Here are the principal file types and what you need to know about them.

Text vs. Binary files

Every file is either in binary format or text format. Text format seems pretty straightforward; when you read files in this format with a text editor, the file contents look like ordinary printed text.

Unfortunately, it's not so easy. When you open a file written for WordPerfect or Word, the contents also look like ordinary text, but these are actually binary files. So what's the difference?

The difference is in the way the text is stored. In text format the set of characters used to describe the data is fairly small; only printable, visible characters are allowed (such as "a," "+," and "%"). In binary format, the files also contain special characters that you can't see. These characters control formatting, fonts, page layout, and more—features not available in the plain text file.

Text files are usually pretty easy to handle because there is not such a large number of different approaches to storing text as for binary files. There is more than one way to do it, unfortunately; see the following discussion.

If the file's not in text format, it must be some form of a binary file. This means that you can't view its contents with a text editor (or not in any meaningful way); you have to do some processing on the file before the contents are useful. This processing might mean displaying it with a word processing program, decompressing it, displaying it with a graphics viewing program, or running a video decompression program on it to display it as a movie. Presumably you know what the file is when you retrieve it (otherwise why bother?) but if not, the following discussion might help you identify it.

Note that binary and text files have some different requirements for file transfer. Some transfer programs, including the UNIX command *ftp*, frequently used in Internet transfers, require you to state whether you are transferring a text or binary file. In other cases, you can't send binary files through e-mail gateways. So though you can send a word processing file to a colleague if you both have accounts on CompuServe, you can't send it to him if he has an account on America Online and yours is on CompuServe. If you run into this restriction, turn to Chapter 14 for a discussion of the *uuencode* and *uudecode* commands, which translate binary files to text files and back again.

Whether the file is text or binary, there are still some compatibility issues facing you before you can use it.

Text Formats

The fact that you have a text file doesn't mean you're free of compatibility issues. While the issues are fewer than with binaries, there are still a couple of tricks.

One is that there are two ways to store characters. Each character in a file is actually represented with a number. But rather than just assigning a=1, b=2, and so on, two different groups developed different representation sets. These sets are called ASCII and EBCDIC.

You'll hardly ever see EBCDIC text files in your on-line roamings. This was the format favored by IBM, and it exists primarily on mainframe IBM systems. Since nearly all of the systems you'll be visiting will be PCs, Macintoshes, or UNIX systems, nearly all of the text files you see will probably be in ASCII format. It's so ubiquitous, in fact, that you don't need to know any more about EBCDIC.

You do need to know more about ASCII text files, though. These three types of systems—Macs, PCs, and UNIX systems—all represent a line feed slightly differently. On the PC, it's represented with two characters: a carriage return and a line feed. On UNIX systems, it's just a line feed, but on Macs it's just a carriage return. So if you get a text file created on a Mac and display it on your PC, all of the lines appear in a pile on top of each other so that all you see is the last one, because there are no line feed characters to tell it to advance the lines. Some smart text editors will figure this out, and may add the missing characters.

To get around this, you need to set a flag in your communications software to either strip the extra character if you're sending from a PC, add the appropriate one if you're receiving from a PC, or replace it if you're going between Mac and UNIX systems. How you go about this depends on the software itself; try looking in the configuration under Line Feed or Carriage Return. Macintosh systems now come with a set of file exchange utilities that let them gracefully handle these details; use one of these if possible.

There is another text compatibility problem coming up, but it is not here yet. That is that the 127-character ASCII set, which uses 7 bits per character, is useful for representing only three languages in the world: English, Hawaiian, and Swahili. Every other language has characters that do not appear in the basic ASCII set. To solve this, the international standards committees developed an extension set several years ago. This 256-character Extended ASCII set adds accented characters such as y, ê, and å, and widens the field of represented languages considerably by representing each character in 8 bits. Most reasonably sophisticated text editors now can handle these extended characters without problems, but some older editors or text handling programs find them very confusing. Not only that, the set is still too small to represent all the languages in the world. To do that, you need 16 bits, and the standards committees are working on this. It's not likely to be a widespread issue soon, but existing text handling software will certainly not be able to handle text represented in these formats.

Compressed Files

Now that you know that your prize new file is a binary file, you need to determine what kind of binary file. Very often, files you download from other systems are compressed. This saves space on the server where the files are stored, and it saves you time (and therefore usually money) when you download files. Each major platform has its own primary compression technique.

The preferred mechanism for compressing files for the Macintosh is an compression utility called StuffIt; files in this format typically end in .SIT. You can get the utility from most of the major on-line services such as CompuServe.

Compressed PC files most often use PKZIP for their compression; files in this format will have the extension .ZIP. Again, you can get the utility by downloading it from a number of places, including the on-line services.

On UNIX systems, most of the compressed files are stored in TAR format. If you're decompressing them on a UNIX system, you can just use the TAR command there, but chances are you're going to transfer them to your Mac or PC first. This is no problem; utilities like WinZip for Windows can uncompress TAR files as well as ZIP files. While you can find software that will uncompress TAR files on PCs, for instance, make sure that makes sense. Unzipping a GIF file for display on your Mac is no problem (as long as you have a GIF viewer); but using TAR to decompress a UNIX program for your PC still won't work.

Interestingly, if you download one of these compression utilities from an on-line service or other source, you'll get it in compressed format, presenting a chicken-and-egg problem. The manufacturers solve this by storing the programs and related files as *self-extracting* compressed files. This means that the compressed data is stored in a file that has a program segment which contains a small-scale version of the program; you run the downloaded file as a program and it extracts the files for you. Many other files available for download are stored this way, so that the material can be used even by people who don't have compression utilities.

Program Files

When you go out hunting for new software to run on your computer, you need to know the type of software file you're getting. Software written and compiled for a PC, for instance, will not run on a Macintosh, and software for UNIX systems won't run on either. Before you download it, make sure it works on your system.

If the software is available as source code, though, you can get it in text format (the program code), and compile it on your own computer to run there. As described earlier in this chapter, this may give you the means to run software designed for one platform on another, but usually this will require a considerable amount of programming skill. At the very least it will require a compiler.

Picture Files

If you need a picture of a seaside scene, or perhaps some pictures of executives, to spruce up your company's annual report, you can probably find something that fits the bill in one of the many stores of graphics files. Several Internet Usenet newsgroups are devoted entirely to picture files, in fact. You should note before you use them, though, that not all of these pictures are public domain. Anyone with a photo, a scanner, and a modem can upload these files to an on-line database, sometimes at risk of violating a copyright. This is an important issue for all manner of on-line material, actually, but picture files seem to be particularly prone to copyright abuse. Caveat emptor.

That issue aside, you still have to figure out how to display the pictures after you download them. Since some 30 picture file formats are in common use, and many more are in occasional use, this is no small task. The file extension may provide some clue; see the chart at the end of the chapter for some of the most common ones.

If you are viewing a lot of picture files, and especially if you want to be able to convert graphics from one format to another, consider getting a display and conversion program such as Hijack PRO from Inset Systems.

Sound and Movie Files

If your system has sound capabilities, you have the wonderful world of WAV files open to you. While this is not the only sound file format available, it is the most common on the PC platform. Another big contender is the MIDI format; both are found in multimedia CD titles.

For the motion video side of things, the best performance comes from MPEG format clips. As with sound files, chances are that you'll find these files on multimedia CDs. If you do find them floating around loose, you'll need video decompression software, and probably hardware, to view them in their full capacity.

Some Common File Extensions

This section presents some of the most common file extensions. If you're not sure what type of file you have downloaded, look at the last three characters in the file's name, or the characters after the period in the name. These abbreviations often tell the format of the file.

If you're on a UNIX system and trying to determine the format of a file, try using the FILE utility, which can tell you with fairly high reliability the type of the file.

extension	description
asm	assembler source code
bmp	windows bitmap graphics file
c	C language source code
cpp	C++ language source code
dib	Windows or OS/2 bitmap graphics file
doc	either text file or Word for Windows file
exe	executable file; could be for the PC or a UNIX system
gif	graphics format popularized by CompuServe
obj	object file. A compiled source file for a PC or UNIX system

extension	description
out	executable for UNIX systems
p	Pascal language source code
pct	Macintosh PICT graphics file
pcx	PC graphics file
sit	compressed file for the Mac StuffIt utility
tar	compressed file for UNIX tar command
tif	tiff graphics file
txt	text file (ASCII)
wav	PC wave sound files
wp, wp6	WordPerfect word processing document
zip	compressed file for PKZIP PC compression utility

Risks

Before you go out and load your computer with software gleaned from bulletin boards, newsgroups, and anonymous servers, you should know that downloading is one of the preferred methods of virus transmission. Bringing a file down to your computer's hard disk will not infect your computer, but as soon as you run the executable (if it is an executable), you give the virus a chance to act, if it's present.

Most of the commercial services such as CompuServe and America Online only make programs available to their user base after they have scanned the software for viruses, so you can be reasonably confident that software downloaded from one of these places is not dangerous to your computer. If you're not sure of the source, though, or just to be safe, get a virus scanning utility at your local computer store, and use it to scan every program you download before you run it.

Summary

This chapter described the wealth of information available in files for downloading to your local computer, and alerted you to some of the compatibility and other issues you need to be aware of to take advantage of this abundance. In Chapter 14 you'll learn some specifics of how to access this information, including how to sift through it all to find what you're looking for.

chapter 9

The Lure of the Internet

The Network of Networks

It's difficult these days to talk about remote access to computers without talking about the Internet, because so much of remote access runs over the Internet or to it in some way. You have probably been hearing a lot about the Internet and the wonderful things it offers—indeed, it would be hard to have missed one of the most hyped and fastest-growing phenomena of the decade. More difficult than hearing about it is sifting through the excitement to get a clear picture of what it offers and how to take advantage of it.

This chapter presents an overview of the wide range of offerings on the Internet, and describes the different types of access. It assumes you know little or nothing about what the Internet comprises or offers aside from the buzzwords, and attempts to fill you in on this fast-growing area of remote connections. If you are already familiar with the offerings and just want information on getting access to them, you'll find what you're looking for in Part III of the book. Chapter 18 describes various connection options and considerations, and Chapter 19 covers search tools, techniques, and tips.

The Parts and the Whole

Much of what is written today about the Internet focuses on the flashiest side of things, the World Wide Web and its multimedia possibilities. Much of the rest of what is written about it covers the apparent disorganization or the complete lack of authority over content.

The glitzy multimedia applications are definitely out there, and this is the fastest-growing part of the Internet. As a result, many services, and the networks they run over, are severely taxed, and performance is often less than stunning. In this chapter and Part III you'll find some hints on maximizing performance. Just a year ago setting up your computer to access the multimedia part of the Internet was a complicated and often frustrating exercise, but since then the market has responded with a number of utilities that set you up automatically.

The claims that the Internet is disorganized and anarchic have some truth to them, but as you learn your way around, you'll find that much of the material available does have a certain organization; the authors and providers of the information don't want it to be difficult to find, after all. Coming from the other side, the constant development of new search tools offers more ways to sift through the huge volume of available data. Administratively, too, the Internet is more organized than it first appears; Usenet in particular operates in a highly democratic fashion, presenting all changes to structure first for discussion, then for vote, to all interested participants.

The more pedestrian, text-based side of using the Internet doesn't get as much attention as it used to, because the graphical interface is so much easier to use. But this chapter presents the basics in terms that apply to both, and discusses the advantages of text-based access; in some cases, forgoing the graphical world for the time being might still be an appropriate choice.

Internet Overview

What Is the Internet?

This is the most-often asked question by people new to the Internet, and gets the most vague answers. In terms of its infrastructure, it is simply a huge network of interconnected networks, all over the world. The backbone of the Internet is a huge data pipeline between major nodes—major population centers on the various continents, usually. From these major nodes, somewhat smaller links form a network of the next tier of cities and research centers. Below that is a network of many more much smaller links that carry from the mid-range points to the individual sites such as your PC. Many of these are temporary links that connect millions of people by telephone line to the Internet. The rest of these links, as well as all of the higher volume connections in the network, are permanently installed dedicated lines of various capacity.

From another point of view, the Internet is the services that run over it—e-mail, Usenet, the World Wide Web. The next section describes these services, giving you an overview of what's available so you can select the access method that best fits what you need. But first, a dab of history will help you understand how the Internet became what it is today.

A Bit of History

The original network that later became the Internet was developed in the 1960s and 1970s as a way to share access to the huge, wildly expensive computers that existed at the time. This was a project of the U.S. Department of Defense's Advanced Research Projects Administration (ARPA), hence the original network's name: ARPANET.

The design of this early network was based on defense strategies, namely, to create a network that was highly resistant to attack. With this in mind, they created a network architecture that routed messages in such a way that if any node was unavailable (because it had been bombed, in the original scenario), the messages would find an alternate route to their destinations.

This simple-sounding architecture has proved to be extremely powerful, and its mark is still apparent in the way the network operates today, though now the reason for a machine being unavailable is probably a hardware or power problem.

Since research and defense were so tightly linked, universities were the next to get on-line, and by 1982 the Internet served more than 200 hosts in over 20 countries around the world. But this network still required expensive dedicated lines because of the protocol it used. The following year, the Department of Defense adopted the TCP/IP protocol that the Internet is still based on today. This change opened the Internet to many more participants, because the new protocol could run over a variety of link types, and could run on any number of different computer types—now including your home or office system.

Who Pays for the Internet?

So who pays for all this? The short answer is: everyone who uses it. If you want a direct Internet feed, you must contact your local telephone provider for the dedicated line appropriate to the amount of data you expect to transmit. Then you arrange for a feed from one of the carriers, such as US Sprint, to provide you with the data. If the Internet were water, this would be analogous to paying for the pipe, according to the size, as well as the volume of water.

As a smaller scale user, you will probably instead contract with someone who has arranged all this, and simply buy access to their Internet-connected system. Many of these arrangements are pay-per-minute, so you pay only for the time on line. Chapter 18 covers payment strategies and structures.

The Services that Make Up the Internet

None of the things that you can do on the Internet *are* the Internet, though many people refer to them as though they were. Nor does having access to them mean you are necessarily *on* the Internet, though for all practical purposes, the distinction is not really important, except to the status conscious.

Rather, the Internet is the group of hundreds of thousands of computers that these services run over and through, and more loosely, the community of people who use these services. Without any one of the services the Internet would still exist, but without the networked computers or the people who make it live, it would wither and vanish.

E-Mail

The introduction to e-mail in Chapter 6 gave you an idea what e-mail offers. Internet e-mail offers all the same features, with the added benefit of the Internet's enormous reach. The Internet e-mail system serves millions of users, and acts as a backbone for many thousands more. With it, you can send e-mail to anyone on the planet, as long as that person has an address that can be written in the Internet format. When you send a message from your America Online (AOL) account, for instance, to a user on CompuServe, you are using the Internet to do it. The CompuServe address might be something like 77777,7777; to translate that to Internet format, you'd address your message to 77777.7777@compuserve.com. Similarly, if the AOL user's ID is DownLink, you'd send your message from the Internet to DownLink@aol.com. Each network such as these that offers an Internet e-mail gateway has its own mechanism for forming Internet-style e-mail addresses; the best source of information is your intended recipient.

People who have accounts on Internet-connected systems have e-mail addresses that somewhat resemble those above, but these addresses have a more organized and well-defined format that uses a system called *domain naming*. In this scheme, the user name appears to the left of the @ symbol, and the domain name of the user's site appears to the right (for example, susi@compsci.caltech.edu). The rightmost element in the domain name indicates the highest level organizational structure of the site (in the example, edu indicates an educational institution), and the division within the site gets more qualified as you move to the left. Outside the U.S., the top level domain name does not describe the type of institution, but rather holds the country code for the country of origin. Some U.S. sites are starting to follow this global convention as well.

There is a great deal more to say about Internet address naming, but you don't need to know all the details to use Internet e-mail. All you need to know is whether an e-mail address is Internet-accessible; the presence or absence of the @ symbol in the address is one indicator, but as described above, you may be able to access a gateway to a non-Internet system by translating the address as necessary.

Internet e-mail takes your office e-mail system out to the world. Many people think that once they have an Internet e-mail address and can thus send and receive messages with anyone else on line, that they are now "on the Internet." From the point of view of their increased access, they're well on the way. But the Internet offers much more, as described in the following sections.

Usenet

Usenet started as an experimental information exchange system between two universities in the United States. The goal was to create a system that would permit participants to post messages on various topics, and respond to previous messages from other participants. The new messages would then get transmitted to all interested parties, and they could then post new comments or responses. The designers wanted a system that was as flexible and automatic as possible.

This rather simple system evolved from this small beginning between two computers dedicated to discussing a couple of topics related to scientific research, to a bulletin board system that covers the world and has over 5,000 different possible topic areas. These areas are called *newsgroups*, and within each newsgroup you find *threads*, which are groups of messages organized around a certain subtopic of the newsgroup's subject. Under the newsgroup alt.caffeine, for instance, you might find threads discussing the caffeine content in various soft drinks, how best to brew coffee, and what tea shop in the Boston area has the best selection of Darjeeling.

The newsgroups themselves are organized, for the most part, into a carefully constructed hierarchical arrangement. Not all sites get all hierarchies, so if you find a newsgroup that interests you, but can't seem to access it from your local system, you may have to experiment and ask your local provider.

Usenet has a culture all its own, and it's strongly recommended that you familiarize yourself with it before you dive in. The group news.newusers has all the information you need; read it for a few days to get a feel for the environment, and to practice using your newsreading software.

File Transfer Capabilities

You'd probably expect that with all these networked computers, getting files from one computer to another would be an important service, and you'd be right. *FTP* (*file transfer protocol*) is a major Internet service, and people use it to exchange documents, send picture files, get software, and deliver work to their employers. Chapter 8 gave an overview of the issues involved in transferring files; Chapter 14 presents a specific example using FTP over the Internet, so that you can see how it works.

E-mail and file transfer capabilities often work together, in two different ways. If your site does not offer access to FTP itself, it is often possible to get files by sending e-mail messages to FTP servers. Such messages must be specifically formatted; Chapter 14 describes how.

On the other side of the equation, it's often possible to use e-mail software to transfer files, if you are sending them from one person to another rather than requesting a resource from an anonymous FTP server. In this case, the success of the operation will depend on the e-mail software that both parties are using, and the gateways through which the message must travel. As described in Chapter 8, many e-mail gateways only allow you to send ASCII files through them, so if you want to send a simple ASCII file to anyone, things are pretty easy. But if you want to send a program or word processing file to a CompuServe user from the Internet (as one example of a gateway that currently only allows ASCII transfers), you must first encode it using UUENCODE, which translates the file to ASCII format, increasing its size by about a third. Then the receiver must then decode it using UUDECODE, which translates it back to binary format so the receiver can use it. Chapter 14 demonstrates this process as well, and lists some sources of the encoding software.

Research Data Sets

An enormous part of the wealth of the Internet is in the form of the amazing variety of research data, maintained by government organizations, research institutions, and private individuals. Some of these are news indexes, raw data from research projects, or the products of efforts to make more information available on-line. For instance, the Louvre museum in Paris offers a database of artworks, and it's possible to access Library of Congress records to search for information about its collection. Chapter 14 offers some suggestions on some of these interesting sources of materials, both on the Internet and available from private sources.

Search Engines

Yes, there's a tremendous amount of fascinating data out there, but finding what you want can be daunting, since the amount of data is huge and there is no single method of organizing it all. Over the years a number of search tools have evolved according to how most sources were organized at various times, and how people were most likely to find them. Nearly all of these tools are still available for you to use, and Chapter 19 is devoted to describing them, how they work, what they are each best suited for, and where to find them.

Interactive Contact

While e-mail is good for private conversations between two (or more) people, and Usenet is perfect for open discussions, sometimes you want a more live feeling to your communications. Enter Talk and IRC (Internet Relay Chat). Both of these are largely appropriate for entertainment rather than business value, so this book does not devote much space to them, but their presence is part of the complete picture of Internet services.

Both Talk and IRC conversations require special software to operate, running on a system connected to the Internet. If you are interested in this live conversational mode of interaction, the book *Internet Access Essentials* in this series covers both features in more detail.

The World Wide Web

More commonly abbreviated WWW or simply the Web, the World Wide Web is not really a separate service of the Internet, but can be better thought of as providing a hypertext, often graphical, interface to the previously completely linear and text-based Internet interface.

Nearly everyone thinks of the Web as a graphical creature, and it's certainly true that its graphical format is where it really shines. But if your system doesn't have the horsepower to handle the graphics, or if your provider doesn't offer powerful enough connections for you to receive the full graphical content, you can still explore the Web in a text based-form. Whether you cruise the Web in graphics or text mode, you'll need a *browser,* a piece of software that interprets the Web's presentation and your commands and lets you navigate the Web. The most common text-based browser is called Lynx, so if you need a text-based interface to the Web, ask your provider if they have Lynx or something like it.

Since most people access the Web with a graphical interface, though, and since that is the way to get the most out of it, the rest of this discussion will focus on that method of access.

An amazing range of people and organizations have presences on the Web in the form of interactive graphical documents called *home pages.* A person's or group's home page is the Web document they have created, with hyperlinks to other documents throughout the Web. When you first access the Web, you will probably start out at the home page of your local provider, and from there you can click on any of the buttons or "hotlinks" on that page, or can enter an address of another page that you are interested in. When you do so, the Web software connects you to another computer that houses the page you have requested. You can chain across the planet this way, linking one server to another, tracking information (the Web is rich in search tools) or just following your fancy. Figure 9.1 shows the home page of the U.S. President's home and office, the White House.

As you can see in the figure, some parts of the screen are highlighted, and others are buttons you can click. This point-and-click mechanism, combined with the possibility for attractive graphics, is what has made the Web so wildly popular. It looks much better than the old text-based interface, and it's much easier to use.

This short textual description of the Web can't possibly do it justice. It's much too interactive and visual for the written word to adequately present it. If you have a chance to try it out, by all means do so. Chapter 19 describes the powerful search tools available on the Web that make it useful, as well as entertaining.

Super High-Powered Applications

At the extreme end of the Internet scale are emerging applications that require tremendous bandwidth, and are currently available only in a few universities and large corporations. These applications run over the MBone, or multicast backbone, which is a virtual network that runs over the Internet. Some of the "programs" that run over the MBone include videos of portions of academic conferences, meetings of the Internet Engineering Task Force, or audiocasts of government hearings. Most of these programs, clearly, are currently of interest primarily to researchers, and available mainly to researchers, and it's unlikely that the average user is going to be able to afford the bandwidth to receive them any-time soon—not to mention the fact that the data intensity of these applications is severely taxing the existing network infrastructure. Thus the network will need to be beefed up considerably before this sort of power is widely available, but it does provide a taste of things to come.

Do You Really Need Internet Access?

Once they get Internet connections and get past the glitz of being on-line, many people use their e-mail capabilities almost to the exclusion of any

Figure 9.1 *The World Wide Web home page of the White House.*

other service. If you're not sure whether you need full access to the Internet, this section describes the main services and some other ways to get access to them.

E-Mail

A great many service providers offer e-mail solutions, and nearly all of them offer gateways to Internet mail. If all you need is e-mail access to the Internet, it might pay to consider alternatives to full Internet access. Check with various on-line providers, as described in Chapter 21. When shopping, keep the following questions in mind: Does the provider charge an extra fee for messages to or from the Internet? Sometimes this charge comes out of a monthly e-mail allowance; you'll have to do some figuring to determine whether your volume is likely to go over this amount. Next question: Does the provider's e-mail gateway allow binary files as e-mail attachments? You may be planning to only send ASCII text right now, but sooner than you think you will need to send a word processor document or graphics file, and you probably will not want to hassle with UUENCODE at that point.

Whether an e-mail-only account makes sense for you depends quite a lot on your local provider situation. Some are so reasonably priced that it's not worth it to bother with less than full Internet access, but this is not the case across the board.

Newsgroups and Their Kin

With just a UUCP account you can get all the newsgroups and post to them without a full Internet connection, though the interface can be cumbersome. As another alternative, many on-line services have their own versions of newsgroups, typically called forums, discussion groups, or bulletin boards. Many of these are moderated, and some are censored. While this doesn't uphold the fullest interpretation of free speech, it can keep the discussions more focused and organized, and prevent people from posting their practice sessions (postings of the form: "does this work?/exit/off///").

Research Databases

There are a number of financial, press, and other databases available on a pay-as-you-go basis, as well as electronic clipping services that will send you articles relevant to your interests. If you need highly specialized data, you need it frequently, and you don't want to have to do a lot of hunting for it, you may be better off subscribing to one of these database services instead. Chapter 14 overviews some of the offerings.

Interactive Gaming and Chatting

Many of the on-line services offer ways to play interactive games and chat on-line, but they may not reach as many people as the Internet.

The World Wide Web

There is no substitute for this one. If you want to get on the Web, you will have to get an Internet connection. Nearly all of the major on-line providers, as well as most of the smaller ones, now offer not only a Web-capable connection, but all the software you need to link your computer to it.

Summary

The Internet is a vast web of interconnected computers and networks, and offers messaging, discussion forums, research tools, and much more. This chapter has provided only the most skeletal of overviews of what the Internet is and offers. For a more comprehensive description of the offerings, including tips on using them effectively, consult *Internet Access Essentials*, also in this series. For specific information on connecting to the Internet or searching using the Internet, turn to Chapters 18 and 19 in this book.

chapter 10

Advanced Messaging

More Than E-Mail

Chapter 6 described e-mail in its most basic form: the ability to send text messages, and perhaps certain kinds of attachments, from your computer to another, whether the other is on your local LAN or on a network halfway across the world.

This chapter explores the further applications of messaging. Advanced messaging is a continuum, starting with more sophisticated ways of handling the simple messages described in Chapter 6, to ways of sending messages between groups of people, to using messaging capabilities to manage group discussions, to the evolution of much smarter messaging systems that use the basic message technology to automate the common tasks of collaborative work. This chapter describes each of these applications and offers some suggestions for taking advantage of them. It wraps up with an overview of the major messaging packages and the strategies of the principal vendors.

Handling Messages Better

Nearly everyone who has been using e-mail for more than a year or two has the same problem: how to deal with the volume of messages that come in every day. Especially in corporate environments, the number of messages in your in box each morning can top 50, depending on your position and corporate culture. Sorting through all of these messages, answering the ones that need answering, storing some for future reference, and deleting the irrelevant ones can take a significant amount of time.

Since many of these messages are of a repetitive nature, the problem is well suited to computer automation. Several companies offer e-mail filters, either integrated into their e-mail client software or as separate standalone packages that work with existing e-mail packages. These filters perform the functions of a secretary and answering machine. For instance, when you are away from the office or have an announcement to make, you can arrange to send out a canned message to everyone who writes to you, explaining when you'll be back or where to get more information about particular topics.

Filters can also examine incoming messages and sort them according to particular criteria that you define. In many office environments, for example, messages announcing absences appear frequently. You could define a filter for your e-mail that examines the subject line of each incoming message for the string "out of the office" and places all messages containing that string in a predefined folder. Furthermore, you can arrange to delete anything in that folder older than a certain age. This way, you always have an up-to-date picture of who is available or not, without having to examine and sort each message personally.

Further filtering capabilities let you sort messages according to sender, so that all messages from certain people could be filed in particular folders. This would be a useful way to manage regular mailings such as status reports or digest forms of discussion lists (see below).

Filtering software can also forward messages according to certain rules. For example, perhaps you recently changed jobs at your publishing firm from Acquisitions Editor to New Technology Editor. You could arrange to have all messages that have to do with acquisitions forwarded directly to the person who now handles that job, freeing you from forwarding all of these messages by hand.

Routing and filtering software, though, works only as well as the rules you define. This can be tricky. Consider the situation where you receive personal and work-related messages over the office e-mail system, and an incoming personal message has a subject line that contains words that cause your sorting filter to interpret it as a message that should be forwarded to someone else—not always what you want.

Mail Robots

So, though they can be powerful tools, e-mail filters and automatic filers have to be used carefully, especially with personal e-mail. Their greatest strength is in automating repetitive responses, such as answers to support questions or common requests. If your work involves a lot of these types of messages, consider using a combination of mail robot and forms server. When a user asks this automated utility for information, he receives an electronic form to fill out with the information that pertains to his question. When he returns the form, the mail robot (sometimes called a *bot*) extracts the various data fields and constructs a database record from the response—useful for gathering marketing information—or selects and sends a particular previously prepared response based on the user's questions.

MAILING LISTS

Distributing Information by E-Mail

The next step up in sophistication from sending canned messages in response to queries is to send messages on a regular basis to a list of subscribers. Interested people send a message to a computer running a program called a list server, and the server adds their e-mail addresses to the subscription list. The next time a message is posted to the list, everyone on the subscription list gets a copy of it as e-mail. Many companies distribute information about new products this way, and the mechanism is also used to disseminate information about technical conferences, concert schedules, and items of particular political interest.

Public Group Discussions

The type of magazine or newsletter format described above is one type of e-mail list service. Another form is the discussion list, where everyone on the list can post messages visible to the rest of the list and can view and respond to messages from the other members. Lists of this type often develop their own cultures and inside jokes. They resemble Usenet newsgroups (see Chapter 9) or forums on the various on-line services (Chapter 21), but often have a more personal atmosphere because they are open only to subscribers. Some of them are available only by invitation; you must prove yourself to have the proper qualifications through your postings to public forums, newsgroups, or lists, before you can join these private lists.

You can receive an enormous selection of mailing lists, both the information type and the discussion variant, if you have access to Internet e-mail. You can find a partial list of the public mailing lists, including descriptions of each list and information on how to subscribe, by anonymous FTP at rtfm.mit.edu under the directory /pub/usenet/news.answers/mail/mailing-lists (chapters 8 and 14 have information on how to get the files that compose this list).

Workgroup Integration

All of the technology that allows you to exchange messages with another person or a group can be expanded to support much more complex message sending between groups, such as document sharing, automation of common group processes like schedule updates, and collaborative decision making. For these applications, the underlying messaging structure supports the sending of the information, and the addition of a shared database for the documents, plus a client/server architecture running over the group's network, supports the body of shared information. The client portion of the applications determine what type of information sharing they facilitate; this section describes some of the current offerings. Some of these applications come as standard parts of existing information sharing systems, others are third-party add-ons, but all are based on expanding the power of messaging technology.

Sharing Documents

Suppose your workgroup is collaborating on a design document for a new project. Each person needs to contribute his or her portion of the design to the finished document. In a simple e-mail system, you'd achieve this by making your changes to the document, then sending it to everyone in the group via e-mail. The members would then be responsible for making sure the copy they are editing is the most recent.

Clearly, this mechanism can lead to some problems, mainly involving merging multiple changes and culling out-of-date documents. In a document-sharing system, you can specify that a particular document is to be shared among certain members, and the system takes care of making sure that all of the changes to the document happen in an organized manner, so that as soon as anyone makes a change, all the members of the group see it. The system also makes sure that multiple changes do not overwrite each other, and handles merging. The result is a system that electronically simulates a group of people sitting in one room, working on a document by hand.

A technique known as replication ensures that this shared information is available not just to people on the same server, but also across the distributed network on other servers. The system updates the information on each server as it changes, so all members see the changes as they happen.

Forms

In many groups, the information that must be collected and distributed can often be codified into a form. The old mechanism for dealing with this type of information, such as customer inquiries, vacation requests, or software change notices, was to use a word processor to create a form, print a stack of blank forms so that the members of the group could take one as needed, have the members fill in the information on the form, then have someone in the group enter the information into a computer.

Instead, forms technology, built into many of the new messaging systems, lets you define a form and make it available to the members of the group. The members can access a form, fill it in with the required information, and send it to the appropriate processing engine. That engine might be the personnel manager, who handles timesheets and vacation requests, or it might be an automated utility that updates the customer database with the new customer's information and schedules a salesperson to call the customer.

Group Discussions over the LAN

Another creative application of messaging technology is software that supports discussions and collaborative decision making, by letting everyone have equal time to put forth their ideas but also letting everyone chime in when they want to. This software simulates an on-line meeting where the participants are not all present at the same time.

Applications of this sort fall into the category of workgroup computing and messaging. The conferencing features resemble forums or newsgroups on on-line services or Usenet. The software organizes the messages into threads, so you can easily see which topic the messages are responding to, and within each topic the messages are organized hierarchically so you can easily see what responses are in answer to which previous postings. Conference+ from MESA is one such product; it is integrated with Microsoft Mail, so you must be using the Microsoft Mail client to use it. Lotus Notes also includes this capability, and it works with many of the major e-mail clients.

Integration with Other Applications

Messaging technology lets the office suite providers add such features as the ability to send scheduling information from a message to the scheduling software, send meeting requests through the e-mail and let recipients automatically add the meeting to their calendars, or perform group presentations over the LAN with the presentation software.

The Underlying Architecture

Messaging, in order to work effectively in a wide range of applications, must operate according to standard defined interfaces. With so much growth potential in advanced messaging solutions, all of the big software companies are anxious to develop a solution that is general and powerful enough that it becomes the architecture of choice for the majority of messaging applications. To this end, three principal groups of companies and organizations have put forth their solutions and are angling for maximum adoption. This section describes what the different solutions offer in terms of strengths and future growth.

Common Mail Call (CMC)

This standard is defined by the X.400 Application Program Interface Association (XAPIA). It's the API of choice for cross-platform development. If you need your messaging system to run on UNIX systems, PC clones, and

Macintoshes, this is the one to go with. Both Microsoft and Lotus support it in their messaging applications, but it doesn't have quite as wide a base as MAPI (described below).

Vendor Independent Messaging (VIM)

This is the name of the consortium that defined the standard, as well as the name of the standard. It's the main competitor to MAPI, and targets the world of Lotus Notes users on Windows and OS/2. It's supported in the Lotus cc:Mail client, but many of the other vendors who originally supported it have moved to supporting CMC instead. It's probably not a good bet for the long run.

Messaging Application Program Interface (MAPI)

MAPI is Microsoft's definition of how messaging should work, and this standard now predominates over the other two, with both Lotus and Novell supporting it. It has two parts: the Simple MAPI (SMAPI) handles the most basic levels of messaging functions, such as sending and receiving; the advanced level, called Extended MAPI (EMAPI) offers modules that let you connect your messaging system to that of some service provider such as CompuServe. These modules are called Service Provider Interfaces (SPIs), and greatly enhance the power of the applications that use them.

Messaging Software

Now that you have a sense of what enhanced messaging capability can offer, this section gives an overview of the major messaging packages currently available.

Lotus Notes from Lotus

Lotus Notes has a reputation for being the most advanced of the messaging systems, and with good reason. It stands a step above the other messaging systems, offering automatic synchronization of changes to shared documents, discussion and decision making support, and planned integration with the Internet. Remote Notes users can access the system with the full interface of the local LAN users; they don't have to sacrifice functionality or work with partial documents or delayed copies of documents.

The key feature of Notes that is missing from the other two main competitors is replication. This important capability, in addition to the large number of third-party applications available for Notes, combine to keep Notes well ahead of the others.

GroupWise from Novell

Since Notes gets all the press attention, it's easy to overlook the other players in the messaging arena. But Novell's entry, GroupWise, boasts twice as many users as Notes, and performs many of the same functions. The focus of GroupWise is more on integration with the Novell suite of office applications than on messaging per se. Novell's strength is in its networking product, NetWare. Since NetWare works on a number of different platforms, Novell has designed a messaging architecture that does the same, and you can find messaging application clients for Windows, DOS, UNIX, and Macintosh platforms.

GroupWise features include scheduling, e-mail, forms, discussions, and calendar applications.

Exchange from Microsoft

At its initial release, Exchange will be available only on Windows 95 clients, but eventually you'll be able to run it on DOS, Macintosh, and UNIX platforms as well. It will include basic messaging capabilities, as well as a forms development package that lets enterprises create their own forms—which they can then share over the Exchange system. You can expect a number of third party applications that run with Exchange clients; Microsoft is offering well-integrated development tools to that end.

Features include seamless support for fax, rich-text, X.400, picture, and other document types; discussion databases that users can view either locally or remotely; and built-in services for schedules, e-mail, and calendar maintenance.

Summary

E-mail is the most common and widely used form of messaging; it represents messaging at its most basic. More advanced messaging technologies move beyond sending simple paragraphs of text between people to less visible but more powerful roles supporting the way people actually work. This chapter describes some of the practical applications that are currently possible using these advanced technologies, from group mailing lists to collaborative work sessions.

p a r t

Implementing Remote Access

III

Making it Work

Part I of the book presented the basic concepts you need to understand remote access. Part II took a look at the various capabilities and applications available to you remotely. Now, in Part III, you'll learn the specifics of how to implement your remote connections.

This section starts off in Chapter 11 with a decision planner. This chapter pulls together the information from the previous two sections so that you can determine quickly what kind of remote access solution is the best for your needs.

Having determined what you want, Chapter 12 takes you shopping. It examines the tradeoffs between what you might want and how much you're willing to spend, and points out some of the practical issues you face when you go to make your purchase.

Once you get the equipment and software home, the next step is to install it in your computer. This is the focus of Chapter 13, which describes ports, interrupts, jumpers, and addresses. It also tells you about some software configuration tricks that can lower your phone bill.

Once you have the configuration up and running, the first thing you might want to do is access the wealth of on-line information. Chapter 14 describes file retrieval in detail, and presents a couple of simple methods that will let you get software and other files no matter how low-tech your system may be.

Chapter 15 presents some more specifics about the options available to you in connecting to another computer and controlling it. You can do this to varying extents with different software packages; this chapter describes how.

If you need to connect to a LAN from a remote location, the information in Chapter 16 will help you select the method that works best according to your LAN and PC combination.

Many people need to use their desktop or portable computers to connect to mainframe computers at least some of the time. The information in Chapter 17 tells you what you need to know to do this, from hardware to software.

If the descriptions of Internet offerings in Chapter 9 convinced you that you need Internet access, Chapter 18 describes the different types. You'll be able to select the appropriate level of access for your skill and interest level.

Once on the Internet, one of the most difficult problems is finding what you're looking for. Chapter 19 describes the principal research tools available on the Internet and explains what sorts of data each one is best suited to finding.

chapter 11

Assessing Your Needs

Before You Buy

The previous two parts of the book presented a survey of the kinds of things you can do with remote access. At this point you probably have a good idea what you want to do. The first step in implementation is to determine what is required for your intended goals. This chapter summarizes the main applications of remote access, and breaks down the equipment considerations according to various criteria, including cost, future growth, and portability. By the end of the chapter you'll have a much clearer idea what equipment you need to achieve your goals.

What Do You Want to Accomplish?

Reviewing the chapters in the last part, you will come up with a list of applications for which you require remote access. To summarize, some of the typical applications are as follows.

- Data-intensive links such as between LANs or to a network
- Video conferencing
- Completely portable links
- E-mail

- Web cruising
- File retrieval
- Remote access to a LAN or mainframe
- Remote control of another computer
- Updating one computer with the directory structure on another
- Participating in newsgroups or forums

Weighing the Factors

With your list of planned applications in mind, you can consider the various constraints on each type of solution. The best solution would run at a speed of at least 56 kbps, be completely wireless, and cost very little. Unfortunately that solution doesn't yet exist, so determining what to get requires making some trade-offs.

To a certain extent, evaluating constraints is a matter of striking a balance. Obviously, if you plan only to send e-mail to and from the office, you don't need a full T1 digital link. On the other hand, if you purchase only what you need today, you may find that you very quickly outgrow your equipment, as you discover new applications that currently don't sound very important or interesting. This section describes each of the constraints and how they affect which applications.

Speed

Your planned applications of remote access determine how much speed you need. Some advisors will tell you to get the fastest possible connection equipment; if cost is no object, that's probably good advice, because you don't have to worry about whether you have room to grow. But considering how fast things change, it might make more sense to get something that covers your needs for now, rather than getting a top-of-the-line setup that becomes obsolete before you have used its full power. Here's what the most common applications require.

Light Usage

Simple e-mail messages or occasional file transfers present the lowest speed requirements. For this type of usage you can even get away with a speed of 2400 bps. The same is true if you are using remote access for infrequent access of newsgroups and forums. This is good news for those who want to use cellular modems or acoustic couplers, two portable solutions that do not boast high speeds. But if you often send large files (over a megabyte) via e-mail, you'll quickly get tired of waiting for them to finish at this rate, and you'll probably want a boost in speed.

With 14.4 kbps modems selling for around US$50, it would be silly to purchase anything slower. But if you have an old 2400 bps modem lying around and to start with you plan only to send e-mail, there's no reason not to make use of it and see how it suits your usage.

Moderate Data Transfers

The more data you plan to send, the more speed you need. For large or frequent file transfers, you'll want a data rate of 9600 bps at the very least. Web cruising is impossible with less than 14.4 kbps because of all that multimedia data that goes over the link; for this application, 28.8 kbps is much preferable. If you spend a lot of time on the Web, you'll soon want an ISDN channel, moving into the digital range.

Data-Intensive Applications

At the next level of data throughput come video conferencing, linking two LANs, and other highly data-intensive applications. If you're gearing up for these types of uses, you'll need at least a BRI digital line, as well as some other reference information, as these applications are beyond the scope of this book.

Location of Use

The easiest remote access setup is to use the equipment in one fixed place, near a reliable phone line and power outlet. If this is not appropriate, things get more complicated. Here's what you need to consider; more tips are offered in Chapter 23.

Highly Portable Modems

In this category you find pocket modems and PC Card (also called PCMCIA) modems. Pocket modems are bigger than PCMCIA, about the size of a pack of cigarettes, and require their own power source, either batteries or a power supply plugged into an electrical outlet. For this reason most people prefer PCMCIA modems. These get their power from the socket in the computer, eliminating one cord from the arrangement, but they have other problems. For one, your computer must have a PCMCIA slot to accommodate them; nearly every portable sold today has at least one PCMCIA slot, though, so this is easy to surmount. More complicated and less predictable is the fact that PCMCIA modems don't work with every type of phone line. If you plug one into a digital phone line, as in a hotel with a digital PBX system, you could destroy the modem. To prevent this, take a line tester with you (see Chapter 23). The catch is that if the tester says you can't use the line with your PCMCIA modem, you'll want a backup, perhaps a pocket modem.

Places Without Easily Accessible Phone Systems

The above discussion of modems assumes you can easily plug your modem cord into the phone system. This is not always the case. In some places the phone is firmly wired to the wall without modular jacks, or you may be traveling to several places with different phone plug configurations. There are a couple of solutions to free you from the tyranny of phone system configurations, including plug adapters, but don't overlook the acoustic coupler. It's a relatively low-tech solution, and operates at a low speed (4800 bps is its top speed, and you can expect much lower on noisy lines), and it is not as light or compact as pocket modems, but you can use it on any pay phone or anywhere you can lift the phone receiver. Again, it's a matter of trade-offs.

Cellular and Radio Modems

It seems that the ideal solution would be to plug a modem into your cellular phone, but data and the U.S. analog cellular network don't cooperate too well, as described in Chapter 4. So unless you are in Europe where the cellular network is already all digital, don't plan on getting a data rate of more than about 4800 bps. A better solution is radio contact, via the RAM Mobile Data or ARDIS networks in the United States. Radio modems are built into popular PDAs such as the Motorola Envoy; access to the radio network is by subscription.

Accessing Nearby Devices

If you want to use remote access to update your notebook and desktop computers to each other every day, you don't need a modem at all. Instead look for systems that incorporate infrared technology. Chapter 15 describes a product that provides the necessary software and hardware in one package, if you don't already have it built into your computer. This technology is also available in printers and other devices.

Cost

If cost is a significant factor in your configuration, or if you already have access equipment that does not represent the state of the art but you'd like to use to get started, don't let yourself be pressured into thinking you must purchase a top-of-the-line configuration in order to accomplish anything.

Again, try to strike a balance; if you know you'll be doing multimedia transfers at some point, but not for the next year, you can very likely get away with a weaker connection solution for now, and reevaluate your needs in a year. This is preferable to having your setup go obsolete before you have the applications to use it fully.

On the other hand, if cost is not a driving factor, the prices of connection equipment such as modems, even 28.8 kbps modems, are very reasonable. If you must buy equipment anyway, and are not absolutely strapped for cash, get the most power you can afford. That way you'll be able to stave off an update purchase longer.

Growth

Though you may currently only plan to use your setup for text e-mail messages and infrequent file transfers, you can probably expect that as you come to rely more on those technologies, you'll want to do more and more. The next step might be to dial into the office LAN and work there as if you were in your office; for this you'll want a heftier link than your old 9600 modem can provide.

Similarly, you may use only a single desktop system now, but as you rely more and more on computing, you'll want to get a portable computer to supplement it, be it a notebook or PDA system. To that end you might want to consider portability issues now when buying a modem for the desktop; you might, for instance, want to get a pocket modem that works with both types of systems.

Looking into the Future

You can't accurately predict exactly what will arrive in the future in the way of remote access technology, but what you can count on is that everything will get smaller, faster, and cheaper. Therefore, if you know you'll want to expand your applications later, but not for a year or more, wait. Buy something now that suits your needs and perhaps one step beyond; don't worry about the three steps down the road, because by the time you get there things will have changed significantly and more options will be available, perhaps more suitable to your needs.

For this reason, some analysts suggest making no purchases in remote access technology unless you can count on a two-year return on your investment.

Summary

This chapter summarized the main applications of remote access and presented the trade-offs and considerations involved in purchasing remote access equipment based on such factors as cost, future growth, and location of use. Now that you have determined from this chapter what you need to acquire to accomplish your goals, the next chapter offers strategies and tips for making a well-informed purchase.

chapter 12

Purchasing the Equipment

Buyer Beware

In the last chapter you determined what types of connection equipment and software you needed based on what types of remote access you plan to do. From there, you're ready to go into a shop and make the necessary purchases.

Depending on what type of connections you want, the purchase phase can be complicated because of the large number of options available. This is especially true of landline modems. Since they've been around for many years and their functionality is fairly stable, there's a wide array of manufacturers and options to choose from. If you're looking for more sophisticated or less standardized solutions, your range of choices will be narrower.

Whatever you're trying to get, it helps to have a checklist in hand when you go shopping, and to prepare a set of questions to ask about the prospective purchase. The better you inform yourself about the options and trade-offs, the more likely you are to be happy with your final purchase.

Warranties and Support

For all items in this chapter, you'll want to take a close look at what types of warranties the manufacturer offers. On a basic modem it should last at least two years, though on a wireless modem or cellular equipment it may be only one year.

Equally important is to find out what type of support the manufacturer provides with the equipment. According to the old economics saw, there's no such thing as a free lunch. The same concept applies to computer equipment of all kinds. It's possible to get a great deal on a high-speed modem that comes with software, but when the software or hardware don't perform as expected, you might discover that the documentation enclosed with the product doesn't contain any information that gives any hint on how to reach the company, or even who they are. Another possibility is that the company offers support, but only for a price, perhaps via a long-distance call that you must pay for, via a set pricing plan per question, or some other method.

One way or another, you end up paying for support, either directly, included in the cost of the equipment, or in your own time and effort as you learn to perform the support functions for yourself. If you want to make the investment in self support it could be very rewarding, but it's helpful to choose it explicitly at the outset.

Before You Go

Before you head for your local computer superstore, make a note of the key features of the system to which you want to add the equipment. You'll want to know the type of system (Macintosh, PC clone, some other variety), the operating system and version level (if you have a PC, note down the version numbers of both DOS and Windows, if you use both), how much memory you have, how big your hard disk is and how much free space it contains, the number and type of ports available if you plan to connect the equipment externally, or the number and type of slots available if you plan to install a board internally. Note the difference too between bus slots and PC Card (PCMCIA) slots, and the fact that PC Card slots come in types: Type 2 and Type 3 are currently the most common. The salesperson might ask you any or all of this; it will be handy to have it written down with you.

If you do plan to install the equipment externally and are not terribly familiar with your computer's ports, make a sketch of the back panel of the computer to take with you. Note which ports have cables plugged into them already, and any captions or diagrams near the ports. Make a similar diagram of the innards if your new equipment belongs inside. Be sure to turn the machine off and open it according to the manufacturer's instructions if you open it up.

A couple of other items will be useful before you head out. First, find a copy of *Computer Shopper* or any of the other large computing magazines. They are filled with ads for mail order equipment companies offering everything from LAN equipment to memory chips. You can find these magazines in the computer section of larger bookstores. The information there will give you a good idea what options are available and what the rock-bottom prices are likely to be. Keep in mind that mail order sources have different overhead costs than walk-in shops, so the prices you find locally may not match those available via mail order. Second, have a pretty good idea how much you want to spend; again, these computer magazines can provide guidelines.

Buying a Modem

Since they have been around for quite a while and their function is pretty well established, the selection and number of options available when purchasing a modem is enormous. For this reason, this chapter concentrates on this most pedestrian of communications devices, to help you navigate through the maze of packages and acronyms.

Packaging

The first thing to determine is the type of packaging you need for your modem, since this may somewhat limit your choices in other areas. The four groups are:

1. Internal card
2. PCMCIA (PC Card)
3. External full size
4. Pocket

Internal Card Modems

This type of modem fits inside your desktop computer; you must have a free slot inside to use it. It gets power from the computer bus, freeing you from worrying about power cords. It may or may not have status lights on the end of the card so you can monitor the operation of your calls; at any rate, unless you have the back of the machine pointed toward you, you'll have a hard time seeing those lights. It may or may not have a speaker; this can be a nice feature for monitoring the call progress.

PC Card, a.k.a. PCMCIA

The slot that this type of device fits into is most common on portables, where space is at a premium, but you can also get them in desktop systems. Not much bigger than credit cards, card modems have connectors on one end, and the rest of the card slides into your PCMCIA slot. The card gets its power from the computer and usually dispenses with niceties such as status lights and speakers.

External Full Size

This variety was much more common a few years ago. Around the size of a pair of paperback books, this device has the disadvantage of requiring its own power cord, cables linking it to the computer's serial port, and space somewhere on your table. It's also somewhat more expensive than the internal variety. It usually has plenty of status lights and a speaker. If you select this variety, make sure the cables come with it.

Pocket

Quite portable, these devices are about the size of a pack of cigarettes. They usually plug into your computer's serial port, and often have lights and speakers. If you plan to travel with it, find out whether the power supply can detect different voltages and how heavy it is. If it can run off of batteries, try to find out how long they last. Some pocket modems are so power hungry that they drain their batteries after just four or five short e-mail checking sessions.

Considering the Trade-offs

Selecting a modem for a desktop system presents the trade-off between using a slot inside the computer for an internal modem and using a port outside for a pocket or full-size external modem. The choice will depend to some extent on whether you have more free slots inside the computer or more free ports outside. It also depends on space and clutter issues; all of those cords can get messy.

If you have a portable computer, you have a different set of considerations. A PCMCIA modem is much smaller than a pocket modem and doesn't take up a serial port, but it's much more sensitive to hot phone lines than the pocket variety, so if you travel a lot, you may find yourself carrying both.

If you have both a portable and a desktop computer and only want to get one modem, a pocket modem is probably your best bet. It travels well, and you can use it in both systems.

Speed and Standards

The next question involves the speed at which the device must run. If your budget is in the area of a couple hundred dollars, it's simple: get the fastest modem on the market. This is probably a 28.8 kbps running the V.34 standard. It's almost unnecessary to specify that it use the Hayes (also called AT) command set, but it can't hurt to check and make sure.

Watch out for fax/modem combinations that have different speeds for the two operations. If the speed is listed as 14400/9600, this means that the fax portion runs at 14.4 kbps, but the modem only at 9600 bps.

Note also whether the quoted speed depends on the modem running in compression mode. If so, you'll get a lower speed when connected to a modem that does not offer the same compression rate.

All modems can automatically adjust downward in speed, so if you'll sometimes be needing to connect to a 2400 bps system, you can still get a fast modem for the rest of the time.

Software

The variety of software offerings is even greater than that of hardware. When buying a modem, if you don't already have communications software in mind, find out if the modem comes with software, and see if you can get a demo of the program's use. Most people opt for third-party communications software, which generally has many more features than the software packages that come with operating systems or modems.

Features

Other features you might want to consider in your modem are fax capability and a speaker. The speaker can be a big help in diagnosing call problems, as the next chapter describes. Note that most modems that offer fax capability let you send faxes only; if you want to receive them with your computer, make sure your fax/modem allows it.

Wireless Setups

If you are buying any kind of wireless solution, you'll have fewer choices in terms of packaging and manufacturer than those buying landline modems. This is because the technology is still relatively new and changing fast, so the field contains only the early entrants. Perhaps in a few years it'll be as varied as that for modems. In the meantime, here are some things to look for.

Cellular Connections

If you already have an analog cellular phone, you can get an interface device that connects between the phone and a standard modem. So-called "smart" interfaces even provide a fake dial tone. The result is a configuration that makes the modem think that it is connecting over land lines. This arrangement is attractive if you already have a cellular phone and a standard modem, but it has some severe limitations in terms of call completion (it may take several tries to connect), reliability (calls are known to drop frequently), and speed (typical is around 2400 bps).

If you have a digital cellular phone the picture is much better; you can get a cellular modem to go with it; look for one with a cellular protocol such as ETC or MNP-10. Though still limited to about 9600 bps, the speed and reliability are much greater than with the analog version. Many people complain that digital cell phones have very poor voice quality; in response, a couple of manufacturers, including Motorola, make phones that can switch between analog and digital operation.

Radio Connections

While not inexpensive, radio connections can be much less costly than cellular, especially if you move around a lot, because the radio networks are nationwide, and they don't assess roaming charges as the cellular networks do. It's possible with cellular data connections to spend as much as $10 to send a fax a few miles away; the radio mechanism is somewhat less extravagant. The costs, however, depend on your provider.

The questions to ask when buying a radio modem are the same as for the others: how it will connect to your computer and what software comes with it? In this case the software question is more important because the hardware requires special software. The other thing you'll want to find out is the cost of using the radio network. Usually there will be a setup fee, a monthly subscription fee, and a charge per minute or possibly by volume of data transferred. If you travel you'll also want to find out what options you'll have with the modem in other countries.

PDAs

The tiny personal digital assistants are designed from the ground up for portability, so they often have radio modems built in. If not, you'll have to find out how to connect one to your PDA. This is an area that is not yet standardized at all, so each offering has a somewhat different way of handling the connection. The vendor should be able to advise you.

If the device does have a radio modem built in, your next questions are the same as for purchasers of radio modems: how does the pricing structure work, what is the coverage of the network, and what options do you have when traveling?

You'll also want to look very carefully at the interoperability between your PDA and your larger computer. Some points to consider:

1. What files can the two exchange? Are formats besides ASCII text possible?
2. Does it connect to the larger system via infrared?
3. If not, it connects via serial cable. Does the cable come with the device?

SUMMARY

Purchasing your connection equipment is variously complicated, depending on the types of devices you select. This chapter described what you need to know to purchase various kinds of modems, including cellular and radio modems, and gave you some questions to consider and prepare before you venture into the shop. The next chapter discusses installing your new purchase.

chapter 13

Installing the Equipment

Setting Up

Now that you have the equipment, the next step is to install it in your computer. With luck, you won't need most of the information in this chapter; some systems require very little user intervention to install new peripheral devices, especially such common ones as modems.

Setting up different types of modems involves an enormous number of variables, and in some cases the procedure is so nonstandard that no general tips are available. Fortunately, in those situations, the manufacturers are aware that the procedure is not part of the average user's repertoire, and they either preinstall and configure everything or provide extensive technical assistance.

In this category fall PDAs, which usually have a radio modem and software bundled together and ready to go. Also very easy to configure are any of the systems from Apple; they have long made a point of ensuring that everything that runs on their systems works well together. At the other end of the scale are cellular variations, which are not yet standardized enough that a general overview of installation is possible. Still, the descriptions in this chapter will be helpful if you are connecting one of these devices to your computer.

Because the number of variables is so large, this chapter provides an overview of the process of installation, rather than a step-by-step detailed walkthrough. With the definitions and tips here, combined with the modem installation instructions and the manual for your computer, you should be able to successfully install and configure the arrangement yourself.

Plug & Play

If you're running a system with Windows 95, and all of the hardware, including the PC itself, boasts the Plug & Play logo, you're in great shape. Installing new hardware should be effortless. These systems run through the entire computer when it boots, quiz each of the devices about what they are and where they're installed, and maintain a file on the hard drive with the information. When you install a new device, you just pop it in the system; the system uses the database to determine the correct addresses and other configuration parameters to get the device to work correctly. It should also install the driver for you. If changes need to be made, the system makes them automatically for you.

It will be several years, though, before everyone has full Plug & Play (PnP) systems; not very many PnP peripherals are yet shipping. In the meantime, most people will have to make do with systems that combine some PnP components with older ones that don't use the technology. If that's your case, you'll need to know just as much about your system as someone without any PnP features, but once you know these details, installing further devices should be easier than for a completely non-PnP system. The things you have to know about are IRQs, ports, and memory addresses; the next section explains.

Peripheral Installation

You don't have to be a computer expert to configure new hardware, but you do have to know a few things about the system. With the coming promise of Plug & Play systems, led by Microsoft and Intel, that may change, as computers get smart enough to configure themselves. Meantime, this section introduces the key ingredients to hardware configuration and installation. You probably will not have to adjust all of the configuration options in this section, especially if you are installing an external device, but you may well have to configure at least one of them.

Interrupts, Ports, and Memory Addresses

The computer communicates with its peripheral devices, such as modems, printers, tape drives, and game attachments, through *ports*. Each device must have its own port in order to function correctly.

It would seem that you should be able simply to plug your devices into distinct positions on the computer and it should figure it out, but unfortunately it's not that easy. Ports are specified by a pair of data items consisting of an address and an *interrupt request line* (*IRQ*), not by physical positions. Thus it's entirely possible to install both your printer and your tape drive on the same port, by specifying that they use the same address and IRQ. Thereafter, every time you tried to print a report, you would end up sending a confusing set of commands to the tape drive.

It may be your job to ensure that this doesn't happen. Modems come configured with port values: a memory address and IRQ combination. In most cases the default values will work fine without your intervention, but sometimes, especially if you are installing the device internally, you may have to change the defaults. The most likely value to need changing is the IRQ, and the manufacturers usually make it fairly easy to do.

At least a couple of the ports on your machine have a general-purpose link to the outside of the system, so that you can plug in a variety of different devices. Serial ports have male connectors of either 9 or 25 pins, and parallel ports present female 25-pin connectors. Figure 13.1 illustrates what to look for. Most newer machines label these connectors with their port names. The most usual will be COM1 and COM2, but you might also see COM3 and COM4. Those latter two ports, though, are usually only used by internal devices.

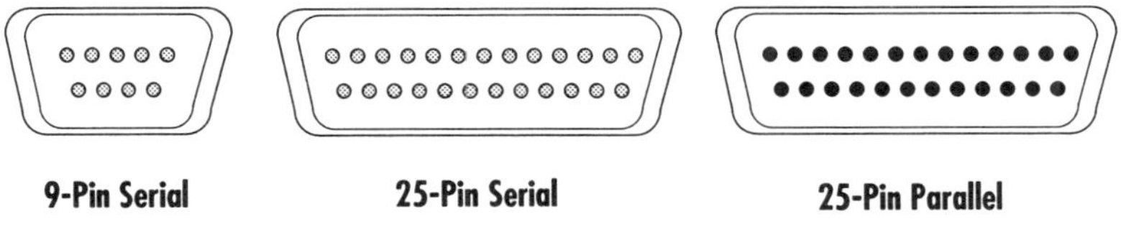

Figure 13.1 *Serial and parallel connectors.*

A number of commercial programs are on the market to simplify all this. These applications aim to identify all of the IRQs in your system, used and unused, and can make the process easier. Check in the utility section of your software shop.

Limited Resources

The problem with IRQs and ports, and the reason you have to know about them, is that there are only a few of them. The PC architecture allows for 15 IRQs in a system, and some of them are reserved or unavailable for certain types of devices. At the same time, most devices can only run on a small number of IRQs. LAN cards, for instance, may give you the choice of using IRQ 3, 4, or 6. If your computer is configured to already use all three of those, you have to find out what the other devices' alternative IRQs are, and set one of those to a different one (perhaps 5), in order to free up an interrupt for the LAN card.

This is the same type of juggling you may need to go through in order to get your modem to work. With luck you won't, but just in case, here's what you need to know.

Internal Installations

If you have a new internal modem, your first step is to put it in the computer. Turn off the machine and open it up according to system instructions. This might be a good time to find the manuals for all of the other devices, too. This type of card becomes its own port, so you may have to juggle interrupts with the existing ports to get it to work.

Before you plug the new card in, take an inventory of the cards already there. Using the manuals that came with the system, note the IRQs and ports used by everything attached to the system. This may take some detective work. It's useful to know that IRQ settings are usually made by moving a small block from one pair of pins on the board to another. Figure 13.2 on page 133 shows such a jumper block, and an example of a diagram indicating what settings it controls.

Now look at your new modem's documentation to find out what settings it uses. If it's configured to use IRQ5, for instance, but another device already uses that interrupt, you'll have to change the interrupt setting according to the documentation. Most modems are shipped configured for

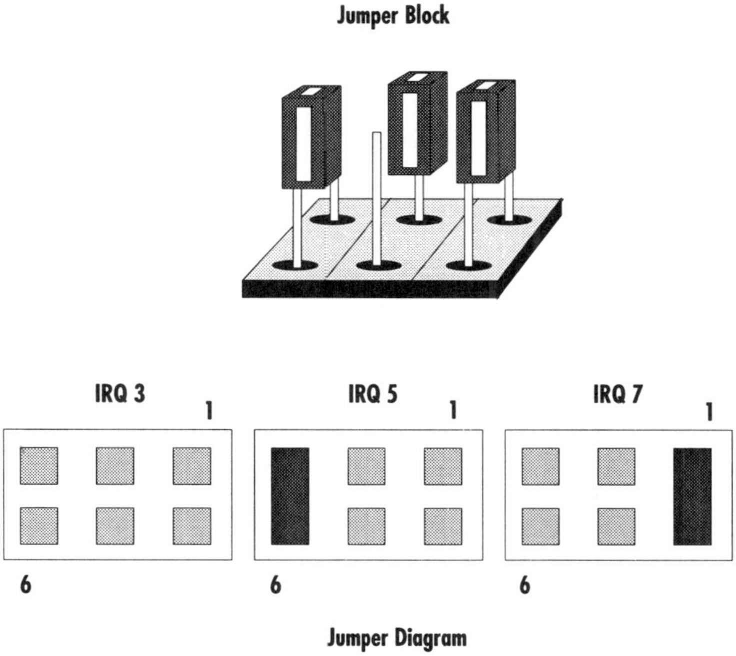

Figure 13.2 *Jumper blocks and jumper diagrams.*

COM2; if nothing else is already using COM2, you're in business. If something is, try setting the modem to COM3 or COM4, but note that you may well run into interrupt conflicts by doing this; go back to your inventory to make sure.

External Installations

With luck, you can just plug your external device into a port. You still have to know what ports are free; there might be some internal device in the system that is already using that port or its interrupt, forcing you to do the same juggling described above. For instance, you sometimes have to disable a particular port in order to use the built-in trackball on a notebook computer, and reenable that port before you connect anything to the external connector. You'd do this by making changes in the system's setup program; see the computer documentation to find out.

PCMCIA Installations

PCMCIA devices, also called PC Card devices, have the happy feature that you can insert and remove them while the computer is running (called *hot swapping*). You still need to install a PCMCIA driver, though, to benefit from this. With the driver running, the system automatically recognizes the card as soon as you insert it. You next need a driver to run the card itself; one will probably be included with the hardware. Still, driver configuration can be a tricky problem. Programs such as CardWizard Pro from SystemSoft Corp. can help, by resolving conflicts between drivers and cards.

Installing Software

Whether the communications software comes with the hardware, or you have selected another package such as ProComm Plus, your next step is to install the software according to directions. The software may prompt you for configuration information during installation, or you may have to configure it after it's installed. Either way, you need to tell it the kind of device you are using and what port you're using it on. Many of the larger software packages have a list of major modem boards from which you can choose.

The First Connection

From your communications software's dialing screen, enter the phone number of a nearby BBS or on-line service node, and the number of data bits, the parity, and number of stop bits. The usual triple here is eight data bits, no parity, one stop bit, represented as 8-N-1. The information that describes how to reach the BBS should include this information. Another common triple, more often found when linking to mainframe computers, is 7-E-1: seven data bits, even parity, and one stop bit. Your communications program will have a space for you to enter these settings as well as the phone number. Now instruct it to connect (or dial, or whatever term it uses), and see what happens.

What should happen is this. You will hear the modem go off hook (as when you pick up the phone), then a dial tone, then dialing just like from your phone. Presently you will hear a high pitched wail that sounds something like a fax machine, then a change of pitch and click as they connect. If not, here are some things to check.

Troubleshooting

If you don't hear anything, make sure your modem has a speaker. If it does not, but it does have a separate jack for an extension phone, plug in a phone to that jack so you can listen to the connection progress.

If you know that the modem has a speaker, try turning up its volume by adding the command ATL3 to the modem's initialization string. Note that loudness commands are not standard parts of the AT command set, so you may have to consult the modem's manual to find out the correct command for your modem to adjust volume.

If the speaker is on, and you hear a click as the modem goes off hook but it's not followed with a dial tone, check the phone cord. Plug it directly from the modem to the wall, and eliminate any devices in between such as answering machines or fax machines, until you get the line to work, then add the extra devices back one at a time, testing after each.

If your phone uses pulse dialing and you hear tones when the modem begins dialing, you need to instruct the modem to dial with pulses. Your communications software may have a setting where you can select either tone or pulse dialing. If not, you must change your modem initialization string again: use the command ATDT for tone dialing, and ATDP for pulse.

If the dialing proceeds without problems and you hear ringing, but no high pitched sound: maybe you're not calling a modem. Check your phone number.

If you hear the connect tone but see no sign of life on the screen, try changing the modem speed to a lower one. If you finally do see something on the screen but it looks like garbage, you're almost there. This problem can nearly always be remedied by changing the parity and bit settings; try changing from 8-N-1 to 7-E-1 or vice versa.

Call Hazards

Even once you're connected, other problems can conspire to make your time on-line more aggravating than pleasant. The first of these is the connection that suddenly drops randomly. Very often, the cause is the Call Waiting feature or some other feature that the phone company uses to notify you that another call is coming in. Modems are not forgiving of these types of interruptions. You can usually cancel Call Waiting before you dial by pressing *66 or some similar code on the telephone's handset. See if it's possible to enter this code into your communications software's dialing list at the front of your phone numbers, or get a separate line for the computer.

Another problem is simply dirty lines. This is harder to solve, and you'll probably need a phone technician to do it. The phone company promises to provide voice grade lines, not data grade, so they may not be terribly sympathetic to your concerns; getting help may take some persuasion. As more and more people use phone lines for computer communications, though (and installing multiple lines to do it!) phone companies are getting more responsive, so if you run into the problem at all, you will probably also get the help you need solving it.

Summary

The process of installing communications equipment can be frustrating and involve lots of trial and error, and the number of considerations is as large as the number of possible system configurations. This chapter presented an overview of the process, and described what you need to know to ensure that the installation goes as smoothly as possible.

chapter 14

Download Strategies

How to Get Those Files Home

Chapter 8 may have whetted your appetite for the types of files that are out there for downloading. It also covered some of the main issues facing you when you go to get those files, including viruses, compatibility issues, and decompression. This chapter provides the practical side of file collecting. It presents a couple of detailed examples, to get you over the bootstrap problem: if you don't have the software, you can't go get the software. With the information in this chapter, you're on your way to being able to get any type of access software available.

This chapter also discusses a few of the major on-line databases. Through these you can get information on prescription drugs, sport techniques, or stock market trends. You can also get customized news clippings delivered daily to your e-mail box. Most of these services are available only for a fee; contact information is in the vendor directory in the appendices.

Getting Files Via E-Mail

Even if the only kind of on-line access you have is via e-mail, you can still get to a lot of files. As long as your e-mail node connects to the Internet, 90 percent of the files available for retrieval on the Internet are within your reach. This means that you can use your e-mail account to get software, data, and information files from all over the world. This is possible because of the existence of a remote control FTP (the Internet File Transfer Protocol) server. You send commands to one of these servers in an e-mail message, and it returns in a day or so with the file you requested. This section describes how to do it.

Text Files

The easiest method of e-mail file retrieval is to get a basic text file from somewhere, because this is what e-mail gateways and systems handle the best. For example, suppose you wanted to get a list of all of the newsgroups. You need to know the server where the file resides, and the directory in which it resides is helpful but not necessary. You also need to know the login ID and password for the FTP server; in the vast majority of the cases, these will be "anonymous" and your e-mail address.

With that information in hand, the next step is to compose your message. Each line of the message must be precisely formatted as follows:

- A line with your e-mail address, so the server knows where to send the file
- A line that lists the server to access
- A line that states whether the transferred file should be sent as ASCII text or in binary format
- A line that specifies the directory on the remote server in which the file resides
- A line that issues the command to get the desired file
- A quit command to finish things up neatly

An example illustrates. The listing in Figure 14.1 on page 139 shows an e-mail message formatted to get the file called newsgroups, in the directory /networking/news/config, on the server ftp.uu.net, and to return the message to the e-mail address 75800.1234@compuserve.com.

Mail the message to ftpmail.decwrl.com, with anything you like in the subject, and after a few hours or days, you'll get the file requested as an e-mail message in your in-box. Now any text file you know of, you can go get via e-mail.

```
reply 75800.1234@compuserve.com
connect ftp.uu.net
ascii
chdir networking/news/config
get newsgroups
quit
```

Figure 14.1 *An automated e-mail FTP request.*

Binary Files, Including Programs

Getting text files as described in the last section is all fine, but a bit limited, because most of the files available on-line are not text files. They may be program files, compressed files, or pictures, but they are seldom data files. So what do you do in the case where you have only e-mail access but want to retrieve some of these binary files?

The answer is UUENCODE and UUDECODE, a pair of programs mentioned in Chapter 8. The former encodes a binary file into ASCII format so that the file can pass through gateways that only admit ASCII transfers. Once it arrives at the destination, the recipient must use UUDECODE to decode it back into binary format.

Unfortunately, there is a chicken and egg problem here, as you can see. You could get the binary programs UUENCODE and UUDECODE via e-mail, but you would have to get them in ASCII format, encoded. You would then need the programs themselves to decode the ASCII format of the program. So if your only Internet access is via e-mail, you'll have to find a copy of UUDECODE somewhere else. Check with your service provider; they very likely make it available to their users. Or check with other users. The program pair is available from all of the major on-line services, and at many Internet sites. If you have a friend with Internet access, have them use Gopher to find the software for you and download it (make sure they download the right version for your system, be it Mac or PC). Then you're back on track.

Now you can use e-mail to get anything at all. By the same remote FTP mechanism shown above, you can also request binary format files, and specify that they be encoded into ASCII using UUENCODE so that they get through the gateways. Figure 14.2 on page 140 illustrates an e-mail message that retrieves an unzip program called gzip111.exe from the site oak.oakland.edu, in a directory called /pub/msdos/compress. It states

```
reply 75800.1234@compuserve.com
connect oak.oakland.edu
binary
chdir /pub/msdos/compress
get gzip111.exe
uuencode
chunksize 50000
quit
```

Figure 14.2 *An automated e-mail FTP request for a binary file.*

that the file is binary, and to transfer it as binary, but to use UUENCODE to encode it to ASCII before it gets to the recipient.

Note the use of something new here: a parameter called *chunksize.* UUENCODed files can get really big, sometimes too big for a single e-mail message. This parameter specifies that if the file is bigger than 50,000 bytes, to break it down into chunks that size—you should adjust for the maximum message size in your e-mail system. UUDECODE will handle the housekeeping involved in putting it back together.

This technique is clearly not the most elegant way to retrieve files from remote sites, and there is a long way to go in terms of inter-computer compatibility issues. But if your only access to the wealth of the Internet is via e-mail, you don't have to be left out of the fun. For a full listing of all of the things you can do via e-mail on the Internet, send a message to this address: mail-server@rtfm.mit.edu. Leave the subject line blank, and create a message body that contains only this line:

```
send pub/usenet/alt.internet.services/Updated_Inter-Network_Mail_Guide
```

Some Notes About FTP

When retrieving files via e-mail as described above, you need to know that the file exists and where it resides. This is not always possible or convenient. Perhaps you have the site name of a server, but a file listing from there that is six months old. The files stored there will certainly be different now from the ones in your listing; you need to get a new listing. You may be able to do this via e-mail. Many FTP servers include a file that lists what files are stored there. This file is often called "ls-lR" on UNIX systems, or perhaps README (UNIX is case sensitive). It usually resides in the /pub directory or the top level directory for the site (the directory name then would be a simple slash, like this: /). So you could try retrieving either of those files via e-mail requests. If that fails to produce results, though, you'll need more interactive access so that you can browse the directories and files available.

Chapter 19 offers some tips on searching for things on the Internet.

On-Line Databases

There are many other sources of on-line information than the Internet, and many of them are designed to be of much more interest to the business user. The Internet, since it was funded primarily by government research and university funds, is still largely of interest to academic researchers. Some of the other on-line databases, on the other hand, are geared specifically toward the user interested in more immediately applicable information. Here's a sample; all contact information is in the appendices.

CompuServe Information Service

One of the biggest on-line service providers, CIS offers access to dozens of data repositories. Some of them are only available through CIS, others offer cheaper rates if you are a CIS member. The other services also offer database access, but none to the extent that you'll find on CompuServe.

For financial and global money information, they offer Global Report. For wider business-related information, including articles from *Forbes* and *Fortune* magazines, Business Database Plus is useful. A trove of health-related information, including information about diet, fitness, and medical data, is available through the Health Database.

Computer information abounds. ZiffNet, the on-line service by the Ziff-Davis publishing company, offers articles and columns from dozens of computing magazines, available within days of when they're published.

For general-interest magazine articles and columns, hundreds are represented in the Magazine Database Plus service. There are also a number of services that will scan the important databases daily and provide you with the newest articles that relate to your specified keywords of interest.

Lexis/Nexis

This huge database, or pair of databases, belong to a company called Mead Data Central. The company has long specialized in selling personalized searches of computer databases on specialized topics, beginning in 1973. Lexis was originally the legal information arm of the service, and Nexis the business information arm, but the two are now merged so you can access them both with one account. Lawyers will find full text listings of case law, plus information on every aspect of insurance, tax, banking, and environmental law in the United States, as well as legal information for much of the rest of the world. On the Nexis side, you'll find the full text of hundreds of magazines and thousands of abstracts.

Dialog

Generally considered the domain of large corporations because of its cost, Dialog has tens of thousands of subscribers all over the world. With careful use, it need not be prohibitively expensive to use. It offers corporate intelligence, sales and marketing information, financial screening information, and a clipping service. It's used mainly by marketing and financial professionals, lawyers, and others who need corporate and financial information.

Dow Jones News and Retrieval Service

This is another important stop for the business user, the investor, or anyone who needs financial data about a particular firm, a market, or about the economy overall. It offers stock indexes, full text versions of the *Wall Street Journal* (including years of back issues), and a clipping service.

Searching Tips

As when searching for information on the Internet, the key to finding what you need is composing the right search string. Unlike on the Internet (usually), on these data services, the on-line clock is ticking the whole time you're on-line, and that can get expensive. To minimize the bill, here are some tips.

- Before you access your service, have a clear idea what you're looking for and jot down a few keywords.
- Know how to form search strings; be familiar with the syntax of the database's search engines.
- Keep your search criteria narrow, but not too narrow.
- Do your thinking off line.

Storing the Information

After you've retrieved just a few important documents or news clippings, you'll discover a new problem: where to put the information so that you can find it again. You can try to impose a directory structure on your data, and discipline yourself to put things in their proper places as soon as you get them. But even if you manage to stick to a system like this, the nature of text documents is that they don't necessarily fit into one particular place or category. And even if they do, finding them can still get tedious as your collection grows.

Several companies are working on software to solve this problem. Some, like PageKeeper from Caere Corp., let you index your electronic documents as well as paper documents that you input using a scanner and character recognition software. From there, the software lets you specify search commands, and its fairly sophisticated searching system rates documents on your drive according to how likely they are to match your search criteria.

Summary

The question in finding electronic information is not whether it's out there, but where. Wading through the enormous quantity of data is a big challenge, requiring a careful approach combined with patience and perseverance. As more and more providers are addressing the issues, the search tools, information available, and means of retrieval are all improving. In the meantime, this chapter provided you with the information to use what's currently available to your best advantage.

chapter 15

Remote PC to PC Options

Levels of Control

If you need to connect one PC to another, you have several options. The appropriate choice depends on the level of control you need to have on the remote computer. Software is available that permits you to simply transfer files from one to the other or between the two computers; to set up the remote computer as a host, letting you move around in its directories and run programs on the remote computer; or to have full remote control, letting you completely control the input devices of the remote computer while your local system acts as its monitor.

This is another area where you benefit from increased competition. The offerings at the lower end of power in this field keep adding features so that the distinction between the different types is blurring. This chapter divides the offerings into these basic areas, but many of them overlap in features and capability.

Transferring Files

If you simply need to transfer files between two computers, and don't need to run programs remotely or do a lot with the files before you transfer them, you have the widest number of options.

If you use both a notebook and a laptop computer, may quickly discover a thorny problem. You take the notebook home to do some work on a report while at home. When you get back into the office in the morning, three people are waiting to see you. You dive into the day, with its round of meetings and appointments, and steal a few minutes here and there to work on the report—on your desktop system. Now you have made changes to an old copy of the report, and have to merge the two versions by hand, a time-consuming and aggravating process.

Several products are available to help. They present on the screen tree representations of the file systems on both computers, and let you drag a file, set of files, or a directory to the other system to initiate an update. In some, you can arrange to have them automatically update only those directories that you prespecify whenever they are in range of each other. They can connect via serial cables, modems, over the LAN, or, most flexible of all, through an infrared port on each computer. So when you walk into the office and set down your laptop and the two computers detect each other, they quietly update so that the newest files are moved to the desktop system. That solves that problem. LapLink for Windows, from Traveling Software, is the most popular of the bunch, and offers a package that includes the infrared hardware along with the software, in case you don't have infrared on your computer yet.

These products are useful for solving other problems as well, including the transfer of files from your old computer to the new one. Often some version of file transfer software comes with your new computer, and they have a smaller, scaled-down version that you can install on the second computer, so that you don't have to buy two licenses for the software, one for each computer.

Host Mode

If you need to access another computer and operate programs on it, perhaps in addition to transferring files, you may already have the software you need. Many communications utilities have a feature that allows you to configure a computer as a host. You leave the machine running in host mode, with the modem listening to the phone line. When you get to your remote destination, you dial into the listening computer, and can operate that computer remotely, moving between directories, transferring files, and perhaps initiating programs. ProComm Plus from Datastorm Technologies is one of the most powerful in this arena, but most communications software packages have the feature.

If you don't have a full-blown communications package, but you do have a fax program (they often come with new PCs), you may be able to upgrade to the communications version for a very reasonable price. In any case, you can get good communications software for well under U.S. $200.

FULL REMOTE CONTROL OPTIONS

The most advanced technology in the arena of remote PC connections is remote control software. The major representatives are pcAnywhere from Symantec, CoSession from Triton Technologies, and CarbonCopy from Microcom, Inc. All give you the ability to connect two computers together and use them as one.

When you dial into a computer running one of these programs, you'll probably first be prompted for a password, depending on the configuration. From there, it's as if you were sitting in front of the remote computer. Your keyboard acts as an extension of the remote computer, so it's as if everything you type had been typed on the remote computer. Similarly, your monitor becomes a window on the remote system, displaying everything that is happening on the other machine.

Common Features

These applications make your screen and keyboard extensions of those on the remote computer. Since you might not want remote passersby to view everything that is happening on the remote screen, your package may offer a privacy option to blank it during the remote session. On the other hand, sometimes you do want a remote user to be able to see what is happening on the remote system, such as when you are collaborating on a document with a remote user. In this mode, the package coordinates the input of both of you into the same application. This is also handy when you both need to work on a project, but the necessary software is only one computer.

In addition to screen blanking, another security feature to consider is dial back capability. That way, you dial into the listening machine, and it hangs up on you. It then calls you back at a prearranged number, ensuring that no one else can use your remote system.

These programs also do all of the things their less elaborate cousins do, including file transfer and host mode control. Some of them include chat utilities so you don't have to combine the remote session with a phone call.

These products are used more and more by technical support personnel, and often come bundled with new PCs or other computer equipment. This enables the tech support person to log directly into the problem machine, view configuration files, and check settings. Doing so eliminates one of the most frustrating things about telephone support, for both customer and technician: the two may speak different languages. With these products, the support person doesn't have to explain to the user what a configuration file is, how to find it, then how to display it; instead, she can just go look at the file herself and get the needed information much more quickly. For support users, the CoSession feature that lets you switch a voice call to data to communicate between the two computers can be especially useful.

Configuration Tips

Note that if you use these products via modem, the phone lines must be pristine. Frequent users report that the biggest difficulty with all of these products is that noisy lines cause the connection to drop, and it can be difficult to reestablish a connection. This requires someone at the remote end to reboot the computer, not always convenient.

Another detail, perhaps more in your control, that is helpful in ensuring performance, is to match the color displays on both systems. It takes a lot of work for the computers to synchronize between 64,000 colors and 16; if the number of colors both systems are running is the same, you may see a significant boost in performance.

Summary

You can find as many different ways of connecting two PCs as you can think of things to do over the connection. Whether you need to simply transfer files between two computers, or move around on the remote computer to find the files to transfer, or fully run applications on the other computer, perhaps in conjunction with the user at the other end of the link, remote communication software is available to fit the bill. This chapter summarized the categories of offerings.

chapter 16

Remote LAN Access

LAN Away from Home

This chapter describes several different technologies that enable you to use the office LAN resources when you're not in the office. With the solutions presented here, you can access the server, printers, and other network services from your home or hotel room.

The remote control software offerings described in the last chapter constitute one solution to the problem. If you want to log into a particular computer that is attached to the LAN, you could install pcAnywhere on that computer, and use that computer to access the rest of the network. This solution is simple and perfectly viable. Remote control options are also available for multiple processors or computers on the LAN, as this chapter describes.

The other solutions in this chapter are designed to connect as a node to the LAN, though, rather than remotely to a particular node. With these types of products, your remote PC becomes a node on the network rather than acting through one. These solutions require changes to the LAN server, in either hardware or software or both, and some require a full dedicated remote access server. Here's what's available and how it works.

Server Node Kits

Nearly all of the major network server providers offer kits that let remote users attach to the LAN as nodes. Among them are Novell with its NetWare Connect, Windows NT with the Remote Access Server (RAS), and Shiva's LANRover products. Some of these require you to add a dedicated remote access server to the network, in addition to the existing server(s) that already form the basis of your network, so they can get expensive. Another potential drawback is that remote users may not be able to access all of the same LAN services as local users, but this is changing fast. Check with the manufacturer of your LAN.

Once you have one of these kits set up on the LAN, with the modems for external users to dial in, you next have to load the LAN driver on your remote computer. If you have Windows 95, this may be very easy; it comes with drivers for all of the remote node solutions listed above. Otherwise, the drivers come with the remote node hardware and software.

Selecting the Right Solution

Implementing a remote node will probably require the assistance of the network administrator. If you are the only one in the organization who needs remote access to the LAN, a modem and remote control software are your best bet, as described above. This mechanism is cost-effective and convenient, but requires some adjustments in thinking, since the file system you are controlling is the one on the remote computer, not on your local system. This does not take long to get used to.

The most economical way to implement this is to just attach a modem to the LAN computer that needs the connection. But if several people in the organization need this type of remote access, consider the multiple processor options provided by companies like Cubix that implement dedicated remote control nodes for the network. These solutions are less expensive than buying the same number of PCs, but operate the same way as dialing into a remote PC and controlling it with remote control software.

Easier to use and more intuitive is the remote node approach, because in that case, you are directly connected to the LAN and can see the LAN drives and access the LAN printers exactly as if you were attached to the LAN in the office. Running this type of connection requires a fairly powerful remote computer that is running networking software (so it also needs somewhat significant quantities of disk space and memory, often both at a premium on a notebook computer).

From the other end, your regular phone line might quickly become an irritating bottleneck if you spend a lot of time accessing a LAN remotely. At that point it's time to seriously consider an ISDN line to your house, and run a PPP, IPX, or other networking protocol over the link. Doing so will speed up your network access significantly; the speed should then be about the same as if you were in the office accessing the LAN.

Simpler Solutions

If the technical and hardware requirements of setting up remote LAN nodes is beyond your interest right now, but you do still have e-mail access to the LAN, there are other possibilities. SoftSolutions 4.1 from Novell comes with a feature called Remote Document Server (RDS). You must have SoftSolutions running on both the server and your local machine to use it. If you meet that requirement, you can use RDS to search for a document on the network via e-mail messages. When found, RDS can mark the document as in use and send it to you for editing. You make the necessary changes to the document on your local system and return the document via e-mail to the server. It updates the document on the server and removes the in-use flag.

Summary

Remote connections to the LAN come in two basic flavors: remote control of a LAN computer, and remote node access to the LAN. This chapter described both and some of the trade-offs between the two, and suggested some vendors to contact for purchasing information.

chapter 17

Terminal Emulation

The Mainframe Connection

A decade or two ago, all computing was done on very large computers. These machines were large in two senses; they had large amounts of resources such as memory and disk space, and they took up large amounts of room.

Now most people work on smaller systems, such as desktop Macintoshes or PCs. Nonetheless, many of the large mainframe systems are still around, and people working on desktop systems still have the need to access the enormous amount of information that is still stored on them.

Since mainframes on the whole appear to be on the way out, providing you with access to a mainframe from your desktop is not considered a growth area, and you're not likely to find a wide variety of products providing you with this type of remote access. Even so, they do exist; this chapter describes what they do, why you need them, what products are available to fill the need, and what you might already have that will fit the bill.

Mainframes: Some Background

If you've done all your computing on a PC or Macintosh desktop type system, it's an adjustment to think of computers with so much power that dozens of people could be working on them at the same time. This is the environment of the mainframe computing model, and it's accomplished by attaching terminals to the large machine. These terminals can perform varying degrees of processing on the data coming from the mainframe, mostly in the form of display processing, and they differ in terms of how many lines, columns, and colors they can display. No matter how powerful they are, the terminals don't come close to rivaling your desktop computer for processing capability, and the vast majority of them don't do much of anything except to display your input and the mainframe's responses. For this reason, they're called "dumb" terminals. The most common ones attach to IBM mainframes and are named by number: 3278 is a monochrome terminal, and the 3279 can display color.

In order to work on a mainframe system in this model, you need a terminal on your desk and a cable running from the terminal to the mainframe itself. Using the terminal's attached keyboard, you enter all your input to the mainframe according to its format, and it responds by reporting back to your terminal the results of your commands. The terminal dutifully displays it.

So much information is still stored on mainframes that this scenario is still an important part of many people's workday, although many large corporations are downsizing their mainframe installations and shifting toward distributed processing and networks based on groups of smaller computers. If you are one of the thousands that need to access mainframe computers but don't want to have a terminal sitting next to your desktop computer, the next section describes how to make your desktop system look to the mainframe like its trusty old terminal.

Simulating the Mainframe Connection

Instead of having a terminal on your desk, it's much more convenient and space-saving to make the PC that's already there behave like a terminal. You need both hardware and software to do it.

Hardware

At the hardware level, you need a board in your computer that performs all of the hardware handshaking and signal exchanges with the mainframe that the terminal was designed to do. The IRMA board by Digital Communications Associates is the most often-used board to do this in PCs, and chances are if you don't use that particular design, you'll end up using something very similar. You then plug the cable lying on the floor, the one that was connected to the terminal, into the port on the emulator board.

Software

Once you have installed the terminal emulation board in your desktop computer, you then need to install software that runs the board, interpreting your input to the mainframe and its responses back to your display. This software is called *terminal emulation software*, and very often it comes bundled with the emulation boards.

Multiple PCs to the Mainframe

If you need to attach several PCs to a mainframe, it might be more economical to put the emulation hardware and software in a single PC and then attach that PC to the LAN; this requires a somewhat different type of emulator board, as your mainframe support personnel can advise. From there, all of the users on the LAN can use terminal emulation software to go through that single PC and link to the mainframe, rather than each having to use up an expansion slot in their computers. This mechanism saves on cabling and emulator board costs, and is especially appropriate if the users need access to the mainframe only occasionally. If many of them need frequent access, though, the traffic might quickly overwhelm the gateway PC.

Remotely Connecting to the Mainframe

The above discussion gives you an overview of what terminal emulation means when the mainframe and the PCs are close enough to each other that you can cable them together, as you would the terminals. But the more interesting question is what to do when they are at remote locations from each other. In this case, you must arrange to attach a modem to the mainframe if one is not already there, and have software running on that modem port that simulates a terminal connection. On your PC, all you have to do is run basic communications software that can dial the modem and emulate a terminal. Most communications software packages come with this capability, and most offer a selection of at least two or three different terminal types. The most common are VT100 and ANSI-BBS. Before you initiate the connection, you'll need to find out what type of emulation the remote computer expects you to be running, and make sure that your communications software offers it.

In Perspective

Actually, there's nothing magical or even unusual about this micro-to-mainframe interface. Although some BBS systems and Internet service providers run their host machine configurations on PCs, when you log onto many of the on-line services such as CompuServe, you are doing nothing more than connecting to a mainframe. The difference is that you very likely have software that the provider offers that hides from you the details of the terminal emulator, leaving you with the substance of the interaction instead. When this happy situation is not the case, you now know what you need to perform the basic access.

Summary

A great many of the computers in the world are large mainframe systems, and though corporations and data banks are downsizing and moving their data stores to distributed networks of linked personal computers, much of the data in the business world still resides on mainframe systems. If you need to access data on one of these, this chapter described the hardware and software you need.

chapter 18

Connecting to the Internet

Types of Internet Connections

Chapter 9 introduced you to the various offerings that you can get via the Internet, so you now have a bit better idea what kinds of things you want to accomplish with your Internet connection. If you have decided that Internet access is a good idea for you, this chapter gives an overview of the different access possibilities, their pros and cons, and what you need to know to make them work for you.

There are several types of access. The simplest type, as far as configuration goes, is basic dial-up, text-based connection. In the most elaborate arrangement, your machine is itself an Internet site, so you have constant, full access to everything the Internet offers. This flavor is the highest-powered option, and also brings with it the most maintenance, responsibility, and expense.

Text or Graphics?

One of the first questions to consider is whether you want a graphically based or a text-based connection. Once you decide that, you may eliminate half of the choices right away.

Graphical Connections

With a graphical connection you can cruise the Web in its full glory, with all the pictures and movies coming up automatically for you, and you can use a mouse to navigate via a point-and-click interface throughout the Web. With the right software (most of which is available via the Web) you can navigate not only the Web itself with this point-and-click ease, but also Usenet, file transfer servers, and most of the other Internet services. This is by far the easiest way to get around the Internet, and it's little surprise that it's growing explosively.

There is a downside to all this, though. To really take full advantage of all these features, you need a relatively high-speed modem—at least 14.4 kbps, and better with 28.8 kbps. Costs of modems continue to plummet, though, so this is no huge barrier. More problematic are the things that are out of your control; the Web is experiencing dramatic growth, and many links and popular sites are simply overwhelmed. You can expect to wait a considerable length of time for graphics, sound files, or movies to transfer to your machine so they can be displayed or played. During peak usage times, you may not be able to reach some sites at all.

To work around some of the performance problems, many Web browsers download the text of a Web page first, so you can read the material there while your browser is busy getting the graphics. If you scroll down the page or connect to another page while a graphic or sound file is downloading, the browser abandons the download process and turns to getting the text for the next page or page segment.

Another feature that is cropping up in browsers is the ability to turn off the automatic downloading of graphics. You can probably do without seeing the company's logo; when you click on a tag that indicates a graphic or sound file you are interested in, your browser goes and fetches only that media file.

Text-Based Connections

This type of connection does not get much attention anymore, because the graphical connection is so attractive and much easier to use, but the majority of Internet access accounts are still of this variety. If you just want to get things done, and don't need a lot of glamour, this is a perfectly serviceable approach to Internet access. The advantages are that you do not need a high-speed modem—your trusty old 9.6 kbps or lower modem will do just fine—and your choice of service providers widens somewhat. A more significant advantage is that the price of the connection may be lower, often considerably lower, for a text-only connection. Chapter 21 describes pricing plans and strategies.

The disadvantage of this type of interface is the learning curve. Instead of pressing buttons and pointing to things with a mouse, you have to learn commands to navigate the system and the Internet. Depending on your provider, you may have to learn commands to get around in UNIX, an operating system that is not noted for its gentleness with beginners. This type of access is also called a UNIX shell account, and it's becoming less common because providers are presenting their users with menus for navigation, rather than a simple UNIX prompt. There are two reasons for this. First of all, system operators and providers don't want to give as much access to their systems that a bare UNIX prompt gives, because of the security risk this presents. Second, these systems are seeing more and more inexperienced users who find UNIX and command-line interfaces too frustrating to use.

Despite the possibly more complicated interface, you don't really sacrifice much by going this route. You still get full hyperlink access to the Web, for instance—as long as your provider supports it—you just don't get to point to the link you want. Instead, you specify it by typing in the number next to it. So if you don't have access to somewhere where you can get a Serial Line Interface Protocol (SLIP) or Point-to-Point Protocol (PPP) connection to the Internet, or your computer is not up to the demands of the software, you don't have to forgo Internet access altogether.

It's Up to You

For many people, the choice between text and graphical Internet access is clear: they only want graphics and have the hardware to handle it. Most providers now support graphical connections in some form, including providing you with client software as needed, so this is a straightforward choice. But if you don't want to pay the extra costs in provider charges, hardware, or on-line time, a text-based connection will serve you perfectly well.

The rest of this section describes the different types of Internet connections and their implications.

Just the E-Mail, Please

If all you need is the ability to exchange e-mail with people other than those on your office LAN or those on the local bulletin board service, you need to find a provider who offers an Internet e-mail gateway. With the simplest shell accounts offering full featured text-based Internet access for under $10 per month in the United States, it's not likely that you'll find a better deal for e-mail only, but it can't hurt to shop around a bit. You may not have to settle for a text-based interface, either. It's possible that the provider may offer graphical software for you to manage your e-mail, or barring that, you can probably find software that does. If you do it yourself, of course, you'll have to install and configure the software on your own.

Dial-Up Non-TCP Connection

This is the variety of access also called a shell account, or UNIX shell account, because the majority of systems that are connected to the Internet and that host these types of accounts are running the UNIX operating system. In this scenario, you use your communications software to dial into a computer that is connected to the Internet, and your terminal emulation software to interact with it. This connected system is maintained by your service provider, who probably will not provide you with software, but who may provide a menu system to make it easier to find your way to the various Internet services mentioned in Chapter 9.

As long as your provider supports them, you can get access to everything on the Internet that people with more elaborate connections get, including the search tools, file transfers, and the Web. What you don't get is the graphical interface, but in exchange you get a cheaper connection (as low as $10 per month for unlimited access in some areas of the United States) and better performance (because you don't have to wait for all those graphics to get transferred to your machine).

The key thing to remember about this type of connection is that you are not directly connected to the Internet; you are removed from it by one level: the link between your computer and the provider's host. This means that when you go to transfer a file from a remote Internet site to your system, you must first retrieve it to the provider's system (using FTP, the search software, or some other Internet utility), and from there download it to your own computer. At first it's easy to forget this extra step, and people can get confused about where their retrieved files are. Refer to the file transfer notes in Chapter 8 for information about converting files to the appropriate formats.

Temporary Full Connections

This is the fastest growing type of connection, for the simple reason that it's so colorful and easy to use (though sometimes not easy to set up). With this type of connection, your local computer is directly connected to the Internet for the duration of the call, then disconnected when you hang up so that the address you were using can be reused by someone else (Internet addresses are a scarce resource these days).

With this type of connection, you not only get all the graphics and ease of use that come with graphical interfaces, you also eliminate the double file transfer problem of the shell account, as described above. You can download files directly from the sites you visit to your own computer in a single step.

As described above, this type of access requires more powerful hardware than a shell account, as well more complex software, and it typically costs more—currently around $30 per month for unlimited access in the United States, though you may find accounts for half that in some areas.

If you want a graphical interface to the Internet, you need this type of connection. To implement it, you need two layers of software: a transport layer (also called a *TCP stack* after the transport protocol it uses) and a user interface layer. This interface layer is usually the Web browser, though it might also be any other Internet utility software, including e-mail applications, FTP software, and so on. This discussion is restricted to Web browsers, because they are the utility most people want to set up, and because they give you access to the other services.

TCP software comes in two flavors: SLIP, the older variety, has some inefficiencies that many people claim are addressed in the newer, faster, PPP.

Most Elaborate: Becoming a Site

If a temporary connection to the Internet is not sufficient, or if you want to create a system that other people can reach from the Internet, it's time to consider becoming a site. If you need a T1 link to the Internet for 60 users, you probably already know it; such configurations are beyond the scope of this book (though Chapter 24 does address some of the security concerns).

This is the most expensive and complicated option. Note that if you want to publish a Web page that other people can access, you don't necessarily have to set up and configure your own Internet site to do it; you may be able to rent space from someone else. Doing this is also beyond the scope of this book, so check with your local service providers for suggestions.

GETTING THE CONNECTION

Once you have determined what type of connection you need, the next step is to find a provider who offers it. One of the most important goals will be to minimize your long-distance charges, so look for one either in your area, or who offers a toll-free access number (also great for traveling), or who has a hub near you. If you live in a major metropolitan area, you'll have a host of choices; people in rural areas will find the selection more limited.

A chicken and egg problem appears here. Most of the collected information about on-line providers is available on-line, so you need access to find out how to get access. There are a couple of ways to get around this. Enlist the help of any friends or acquaintances who have connections; not only can they retrieve the local listings for you, they can tell you about their experiences with their providers—valuable personal experience. Local computer shops may have information, too. Call and ask them if they can name some providers in your area (it will help if you have a clear idea of what type of connection you need). Finally, the trusty Yellow Pages may help. Try looking under Networks, Telecommunications, or Internet.

Chapter 21 discusses some of the main on-line providers in the United States, and offers some suggestions about access through the local operators.

Can You Get It at Work?

Your workplace may already have Internet access. Check with your network administrator to see if they do, and to find out whether you can access it via the in-house LAN. At the same time, you might also want to familiarize yourself with the workplace rules about personal use of the company equipment and network. Usually Internet access is provided for company-related research and projects, and management has been known to take a very dim view of people using it for social or entertainment purposes of any form (despite the temptations).

Configuring Your Connection

When you get your account access information (user ID, access phone number, and so forth), they will probably also give you some tips on setting up the software. Exactly what you need to do depends very much on your computer and type of connection, but this section gives you an overview of what is involved according to the type of access.

For a Text-Based, or Shell, Account

Previous chapters have already given you the information you need for this, the most basic type of access. For it, all you need is a modem (even as low as 4800 bps may be quite sufficient), communications software, and terminal emulation software (the latter two usually come together in most of the more powerful communications software packages). Your provider will tell you the settings for the following: terminal emulation type (often VT100), data speed, parity, and stop bits, all of which you enter into your communications software settings. In addition they will provide you with a user ID code, and an access phone number. They may let you choose your own password, or they may equip you with one to start with; in this case the first thing you should do on-line is change it. Chapter 24 discusses password security in detail.

For a TCP-Based Connection

Your basic communications software does not know enough about Web format files to interpret them and present them to you, nor does it know how to handle the higher speed data protocols demanded by this type of connection. Thus, for this connection, you need the following: a higher speed modem (14.4 or 28.8 kbps), the TCP software, and the browser.

Again, acquiring and configuring this software depends almost entirely on your computer arrangement. Macintosh 7.5 and OS/2 Warp users have one big advantage here: their operating systems come with the transport software built in. Windows users have the advantage that their sheer numbers mean that software for that platform is widely available, and there are a number of variations to choose from. They will find, however, that without at least a 486 with 8 MB of RAM, performance will be painfully slow.

More and more, the providers are having to get in the act here, and some are providing relatively complete packages of transport software and browsers to their subscribers. It all depends on your provider, though; this is something to research before you sign up.

Via a National Provider

Each of the major providers profiled in Chapter 21 provides you with both the transport layer and browser software when you sign up, and offers support in helping you get it running. More and more, these software packages are largely preconfigured so that you don't have to go through the painful hassle of setting all the settings and switches of a year or two ago. Often you can be up and running and on the Web within 30 minutes of getting the software.

Getting the software can be the bottleneck, depending on how you decide to do it. One way to get access to these services is to send in a registration card with your name, address, and credit information; after about two weeks you receive a user ID, password, and diskette with access software. But there is probably a faster way. Most of the major services allow you to register on-line; you just have to find out what the local access number is (as well as the correct speed, parity, and stop bit settings), dial up, and provide your name, address, and credit information that way. From there, you may be able to access a library that contains all of the software you need for your TCP connection, download it, and install it just as if you had the diskette. Not all the big services offer this type of access, but if you want to save some time at the risk of a more involved installation procedure, it's worth looking into.

Either way, at some point you will be happily accessing the Web.

Learning Your Way Around

Once you make it onto the Internet, here are a few key resources to help you get your bearings, learn where to find the most interesting and valuable resources, and get a feeling for how the whole thing works.

Newsgroups

The news.newusers.questions newsgroup is an often-overlooked resource. Some users skip it altogether, figuring they can work out their problems on their own. Often they can, but the information available here can shorten the startup time dramatically. A dedicated group of volunteers spends time here, answering all manner of questions from the "where do I find...?" variety, to how to navigate Usenet and the Internet, to enlightening users on the cultural points of the Usenet.

If you're looking for help setting up or using your system software, this is not the place to ask, though someone here will probably give you some useful pointers on where you might fruitfully ask your question instead. (Tip: The first place they will suggest is with your access provider.) This group is dedicated to questions about using Usenet, and is especially designed to provide a good learning environment for those who are new to the culture and expanse of the Usenet, and Internet, offerings. Spend at least a few days here reading the traffic before posting anywhere; it could save you a lot of hassle.

In news.newusers.announce you'll find postings of more general interest items; these are not just for new users, by the way. The group alt.answers receives postings of a large number of FAQ (frequently asked question) files.

FTP Sites

One of the most useful FTP sites for the beginner has to be at rtfm.mit.edu, where all of the FAQ files are stored. Also stored here is the file that lists all of the newsgroups currently available, as demonstrated in Chapter 14; this file is updated a couple of times a month.

Another useful resource is the software archive at ftp.cica.indiana.edu. Here you can find TCP client software for just about every application and system type you might need.

Web Sites

If you'd rather get your FAQ files off the Web, head to http://www.cis.ohio-state.edu:80/hypertext/faq/usenet. For a rich source of Windows-compatible shareware, try http://coyote.csusm.edu/cwis/winworld/winworld.

html. And for the same wealth for the Macintosh, software is available at http://web.nexor.co.uk/public/mac/archive/welcome.html

Summary

The main choice facing you when you select your Internet access are whether you need a graphical connection or can make do with text. Having decided that, you can find a provider who offers the services you need, and enlist their help if necessary in configuring the software. From there, learning your basic way around the Internet is the next step. Having done that, you're ready to start some serious searches; the next chapter offers some tips.

chapter 19

Research over the Internet

Finding Out

The often-heard statement that anything you want is available on the Internet becomes tiresome quickly once you try to go out there and find something in particular. The aspects of the Internet that make it so unique and interesting, namely that it has no central organization and is composed of such a variety of ingredients, also are the features that make it very hard to find anything, because it has no central indexing or authority responsible for consistency or ease of use.

The topic of searching for information on the Internet could fill its own book, and there are some good ones out there. This chapter does not attempt to cover the topic exhaustively, but rather to introduce you to the basics of searching for things on the Internet, from software to source code to data repositories. It presents the principal search engines, describes how they work, and explains what you need to know to find them.

The Major Search Engines

The evolution of the Internet search tools parallels the development of the Internet itself, and understanding what each tool was developed for will help you determine which ones are most appropriate for your purposes. This section describes the major tools, how they came into being, what they index, and where you can find them.

Archie

The name is a loose abbreviation of the word "archive." Archie was developed by a researcher who needed to know where on the Internet he could find archives of software available for FTP retrieval. At first he would periodically connect to each computer he knew stored software, make a copy of the list and descriptions of the items stored there, and compare the list to the previous list from that site, noting what was new. He soon automated the procedure, and used various UNIX text searching utilities to scan for keywords to locate particular files. This was the birth of archie. The utility has since then grown into a powerful automated search engine of Internet sites offering much more than just software, and indexing all manner of data sets, documents, FTP indexes, and software storage facilities. If you can get it via FTP, you may well be able to find it with archie.

To some extent, your success when searching with archie depends on the description of files provided by the people who have made their data available, because archie relies on those descriptions when performing keyword searches. Because the file names are also listed in the description files, you can also use archie to search for files by name, if you happen to know the name but not the address.

Archie is implemented on a client/server model, so you need two things to use it: archie client software on your local computer (or access to a computer that does have the client software), and access to an archie server. There are a few public archie servers, and they are often very heavily used (and therefore slow).

Despite these drawbacks, archie is a powerful tool if you are looking for a specific data set or software program among the world's vast TCP stores.

Gopher

This is one of the earliest utilities developed for trying to tame the complexity of navigating the Internet. As with many of the Internet tools, the name of the service is a pun; this one refers to specifying what you want so the service can "go for" it.

Gopher is not a search tool per se, but rather a client/server menu system designed to present a uniform interface between the variously organized Internet data repositories. When you run a Gopher client on your local computer, and connect to a Gopher server somewhere on the Internet, you get a menu listing the resources available on that server. Some of the resources will be menu items that offer something like "Other Gopher and Information Servers." If you select one of these items, the Gopher software invisibly connects you to the server that menu item links to; you don't have to know anything about how to connect to that system. This is the power of Gopher: you can move between Internet sites without having to know anything about the details of accessing each system. Similarly, if you find a resource you want during a Gopher hunt, you can retrieve it to your local system using a Gopher command, without having to type in the FTP command syntax.

Though Gopher is a navigation tool and not a search tool, there are other Internet search tools that use it, so it's useful to know about it. It's also a handy way to cruise around the Internet if you don't happen to have World Wide Web access. Graphical clients for Gopher are now available for most common hardware platforms; consult the software repositories listed in Chapter 18.

Veronica

As archie is to FTP repositories, veronica is to Gopher servers. This is a tool that searches for specified keywords in the menus of Gopher servers around the world. When you perform a search on a veronica server, you enter one or more keywords—multiple searches are supported, as are Boolean expressions—and the server goes out to all of the Gopher servers it can find, looking for your keywords or expression within the text of the Gopher menu entries in each server it visits. This is can be a powerful search method, though it tends to find an unmanageable number of "matches" that you then have to wade through.

You can access veronica from most Gopher servers (as well as from the Web, described below), so all you need is the Gopher access, rather than another set of software explicitly to run veronica.

WAIS

The search tools discussed thus far search only descriptions of archives or of menu items, and so rely on the quality of the file name or description provided by the people who made the material available. If you don't know the exact file name, though, or can't find any matches using veronica queries, the next step up in power is using a *WAIS* (*wide area information service*) search. It searches based on your query in not only names and descriptions of files, but in the files themselves. This is clearly a staggering amount of information, and it's only in the last decade or so that enough computing power could be spared for such a task.

WAIS searches are very smart. You can specify a number of keywords and trigger the search, and after a time (sometimes a long time), you'll be presented with a list of potential matches. You can select promising items to download, or further refine the search. WAIS searches use a technique called *relevance matching,* which lets you refine according to certain items; in effect, you tell it "I want to concentrate on those items that are like these."

Since WAIS searches are so much more thorough than those offered by the other tools mentioned above, their servers tend to be heavily utilized. Aim for off-peak times for best performance—in particular, avoid early evenings on weekdays.

The World Wide Web

Most people don't think of the Web as a search engine, but it started out as one. The developers wanted to present a way to navigate through the Internet not via the menu-oriented structure of Gopher, but rather though hyperlinks, so that a user could follow the paths of his interests, rather than being forced down the paths of the menu designers' visions.

Now the Web is thought of more in terms of its graphical hyperlinked face that presents on-line magazines, advertisements, music videos, and more. But you can also use the Web to search for materials, or more specifically, to access search engines. It offers access to a wide range of different search engines; selecting the right one will depend on your application (what you want to search for) and your access type (how you are connected to the Web).

One such Web-accessible search engine is called Sleuth. It provides quick keyword searches of online databases, and is a good choice if you know the material you want is in an indexed database. Try it at http://www.intbc.com/sleuth.

Another way to search is via the hierarchically organized index provided by Yahoo. This search mechanism is very easy to use, but is very heavily utilized, and accessing the various documents can take quite a long time. For this reason, you probably won't want to rely on it if you have to pay for your connection by the minute. In that case, opt for a keyword search instead, or stay up until the wee hours when rates go down and fewer people are abroad. You can try Yahoo at http://www.yahoo.com.

Yet another possibility is the Web Crawler. It indexes hundreds of thousands of Web documents, and performs fairly quick and well-focused searches. Access it at http://webcrawler.cs.washington.edu/WebCrawler/WebQuery.html.

There are dozens of other ways to search via the Web, far too many to list here. One of the best things about the Web is that it is very easy to find the search tools. Head for any of the sites listed here to start your search, and you'll find other possibilities there.

Usenet

No, Usenet is not a search tool, and it is not often accused of being anything other than chaotic and disorganized. But don't overlook it as a information mine. As demonstrated in Chapter 14, you can get the list of newsgroups from a number of sources; once you have it, you can perform a simple search yourself for newsgroups that might be relevant to what you want to find. Having found the group, subscribe to it and read the traffic for a while, to see if the content really is related to what you seek. Find the FAQ (an FTP site is listed in Chapter 18) and read through it—these files are often rich with resources and pointers to other sources of information.

If your provider doesn't carry a newsgroup that you need, ask if they can add it to the list. Many providers, especially the smaller ones, keep the list down because the volume is so huge, but willingly add specific groups that users request.

If your initial search for a relevant newsgroup is not fruitful, there is actually a newsgroup dedicated to helping people find newsgroups; you can post a message there. The group is called news.groups.questions.

Where to Start

As you can see from the above overview of Internet tools and sources, finding materials requires you to familiarize yourself rather thoroughly with the available search methods in order to have any kind of success. A lot of research is going into making this process much more simpler and transparent, but success is still down the road. In the meantime, your best strategies are as follows.

Know the Search Engine

One of the first tricks you will learn is that different search engines are more appropriate for different tasks. If you are looking for Mosaic software, for instance, a Gopher search with the keyword "mosaic" will probably be fruitful, because the software files are usually named with the name of the software. But if you're looking for a document describing Mosaic, you'd do better to head to the FAQ archives and search there.

Just as some locating tools are better at finding software and others at literary references, some Web search tools are better suited for certain types of queries. Suppose you are searching for information related to the recreational or educational uses of government-owned forest land. You might want to start your search with a look at a hierarchically organized search tool, to find what you want under Forestry, Agriculture, or Environment, or you might want to just dive into a keyword-based search engine with the same keywords. In the former case, though, you have a better idea which avenues are more profitable as you go. The overview of each search engine in this chapter will help you choose the best one for each task, but getting familiar with them is much more important.

Know how each engine processes keywords. Usually they will make clear when they prompt you for the search string whether they permit multiple keywords or Boolean expressions, and how you should separate each word in the query. But make sure you know the rules; an empty search could be because of a missing comma rather than a lack of data.

Narrow the Search

Before you start, define what it is exactly that you need to find out. Just saying you want to know about the aloe vera plant is much too broad; you may end up sifting through hundreds of matches. Try to narrow the topic down to cultivation, uses for the fiber, or medicinal uses. Again, it takes some experience to form effective search strings, but a little planning can go a long way toward saving you time on line.

The Future of Internet Searching

A lot of research is going into trying to tame the massive quantity of stuff available on the Internet, and make it easier to locate information. As the amount of data continues to mushroom, current search technologies that rely on keywords, or even on Boolean expressions, are fast becoming inadequate. Many tools determine relevancy based on the frequency of appearance of a desired string; this just turns up too many possible matches. Current research focuses on linguistic analysis of text to search for meaning within documents instead. Rather than flat matching searches, one new approach is to try to identify words that are both directly and indirectly related to the words you specify in your search. A drawback is that the documents to be searched must be indexed according to related keywords; software is in the works to automate that process as well.

This sort of analysis is extremely demanding of processing power, and so is only becoming available with the latest wave of high-performance processors, though it's been under development in various forms for decades. Though it's still too computer-intensive for your desktop system, in the next couple of years you can probably expect to find on-line servers that permit searches based on these technologies.

Summary

The Internet makes available far more information than anyone can digest, but finding what you might want can be a complex task. You can access a number of tools and search utilities to automate the task; the success of the search depends on the appropriateness of the tool and your familiarity with it. This chapter described the main tools and gave you some suggestions on where to start and how to increase the odds of finding what you're seeking.

part

IV

Keeping It Useful

Day to Day Remote Access

In previous parts of the book, you learned the basic language and concepts of remote access, what applications are available to you through remote access, and how to set up your own remote access configuration. This part wraps it all up with tips and information on applying your remote access capabilities.

The first chapter of this section, Chapter 20, serves as a checklist for all of the tips offered throughout the book. It pulls together the recommendations on portable computers, e-mail software, connection equipment, and more, so when you go shopping for any aspect of remote access technology, you have a checklist for quick reference.

Chapter 21 describes the major U.S. on-line service providers, and compares their features, offerings, and atmospheres. When you decide you need access to a remote service provider, this section will give you the background you need to choose among the possibilities.

If you are considering making remote work a large part of your work life, you may find the tips in Chapter 22 useful. This chapter is geared toward the professional who is either working entirely or primarily away from an office, and outlines the main issues and caveats affecting this kind of work.

When you take your computer on the road, whether inside the U.S. or out, Chapter 23 presents what you need to know to ensure that you can still connect to your remote resources.

Finally, Chapter 24 wraps up with a description of security. Some people liken the new and fast-growing realm of remote connections to the Wild West: lawless and dangerous. Use the information and resources in this chapter to keep your risk down.

chapter 20

Remote Access Shopping Checklist

Quick Reference

Throughout the book you have been getting tips, advice, and notes about what to consider, what to avoid, and what features to look for in remote access hardware, software, and service. This chapter brings all of those hints together into one place for quick reference.

Modems

- Speeds: Get at least 14.4 kbps if you can possibly afford it; 28.8 kbps will last you longer.
- Fax/Modems: Make sure the speeds of the modem and fax capabilities match, or that you realize your modem speed is slower than the fax speed.
- Fax capability: Look for send and receive; many are send capable only.
- Standards: Look for V.34, with the Hayes (a.k.a. AT) command set.
- Speaker: Nice to have, but rare on PC Card (PCMCIA) modems.

Portability Features

- Smallest and lightest are PC Card (PCMCIA) modems.
- PC Card modems can be damaged by "hot" phone lines; take a line tester.
- Pocket modems: Find out how they get AC power, how long the batteries last.

Cellular Modems

- If you already have an analog cell phone and want to use it with a conventional modem; get an interface between them that can simulate a dial tone; this makes connections easier.
- Better: Buy a cellular modem with a cellular protocol built in. Look for ETC or MNP-10 error correction.
- If you are buying the cell phone: consider analog digital switchable phones. You'll get better voice quality over analog, and better data performance over digital.

PDAs

- File compatibility: It should be able to exchange at least ASCII text with your desktop system.
- Application compatibility: Look for the ability to exchange data with your desktop finance, scheduling, address book software.
- Infrared: IR ports should conform to the IrDA standard.
- Wireless communication connection: radio is preferable to cellular, especially for travel.

Computers that Travel

- Low battery indicator: meter is preferable to a beep.

- Power supply: switchable between 220 and 110, preferably automatic.
- Batteries: lithium ion are the lightest per hour use, but somewhat rare yet.
- Power cord: Modular power cord so you can swap it for the local variety.
- PC Card modem.
- Line tester to protect the card modem.
- Pocket modem in case the line tester indicates that the line will destroy your PC Card modem.
- Acoustic coupler for communication, depending on where you're traveling.

Communications Software

- Setup: Know how to change your initialization string.
- Maintenance: Look for easy management of dialing number lists.
- Reconfiguration: It should make it easy to change the dialing settings to accommodate a prefix for hotels and PBXs.
- E-mail: Look for integrated message management software.
- If upgrading: Make sure it has an import feature to convert any existing data; you don't want to type all of those site numbers in again.
- Fax features: Fax send/receive capability can be very useful.
- Mainframes: If you will need to connect to a mainframe, make sure the software offers the right terminal emulation.

E-Mail Software

Features

- Attachments: The ability to send graphics and sounds, along with messages.

- Gateways to important services. Can you send binaries through the gateway?
- Quoting from a previous message in the reply.
- Fax: Some systems make it as easy to send a fax from your desktop as an e-mail message.
- Encryption for sensitive messages.

Managing Messages

- Sorting: Should be able to view by a number of criteria, including date, sender.
- Filtering: To specify special handling of messages that have certain characteristics.
- Routing: Forwarding, copying messages to other users.
- Filing: A means to store messages according to your own system.
- Archiving: Storing messages long term. Can you delete messages from archives?

Managing Addresses

- Organization: Note how it lets you organize your addresses, search for them, add, change, and delete them, include comments with them.
- Mailing lists: Can you create your own? Does the system allow public ones that all users can use?
- Import from old e-mail system: Critical capability if you have old e-mail addresses that you want to use in the new system.

Internet Service Providers

Services

- E-mail: Can you send binaries across the gateway from the service to the Internet?

- Extra charge for Internet-based e-mail?
- FTP: File Transfer Protocol. Useful for getting data and software.
- Telnet: Remote login capability. Usually only useful if you have an account on the remote system.
- WWW: Look for a service that provides graphical access to the World Wide Web, preferably via PPP.
- Interface: UNIX command line, menus, or graphical interface.

Payment Variations

- Flat fee per month, unlimited time, is the best deal.
- Watch for charges based on volume transferred, volume stored, time on line, speed of connection, or other variables.

SUMMARY

Making the best decision when selecting your on-line service, purchasing remote connection equipment, or choosing communications software, involves making a lot of trade-offs between what you want to accomplish, what you can spend, and what's available. This chapter summarized the main advisory hints from the rest of the book on how to purchase such items as cellular modems, conventional modems, and Internet access, so that you have a one-stop checklist of the most important trade-offs.

chapter 21

On-Line Service Providers

Getting Access

If at this point you have determined that you need a wider connection than just to the office e-mail system or LAN, or just to a single other computer, then your next step will be to select an on-line service provider to supply you with access.

On-line services have changed considerably in recent years, largely because of the sudden growth of the Internet. In the past, these services provided discussion groups for their members, business databases and investor information sources, and hobby and recreation resources. With the Internet attracting so many people, they have tightened their focus, expanded their offerings to compete, and in most cases, added Internet access to their plates of services. The consumer benefits from the resulting variety. If you opt to get your Internet access from one of these providers, you also get access, in many cases, to well-developed communities with more friendly atmospheres than most areas of Internet exchange.

This chapter describes the most common access sources, and compares the major U.S. providers. It also discusses the various pricing strategies you might encounter, so you will know what questions to ask when you compare offerings.

The Major U.S. Providers

All of the services described here offer their own discussion boards, e-mail, and file centers, as well as varying degrees of Internet access. Each offers the ability to send e-mail to and from the Internet, though most do not allow binary attachments.

Providers such as these can be by far the easiest to access, and most provide their own graphical software that you can use to access their service and sometimes the Internet as well. The installation of this software is usually much more automated than what you are likely to get from a smaller provider, but this type of access is also likely to be the slowest and most expensive way of getting on the Internet. They all offer the advantage of being accessible via either local phone numbers or toll-free numbers from most U.S. cities, so if you are traveling, they might offer a convenient way to stay in touch via e-mail. All offer at least 14.4 kbps access, and most expect to offer at least some 28.8 kbps nodes by Fall 1995.

This survey only describes those services that had been in operation for at least a year as of Spring 1995. Several more are likely to come on line by the end of 1995. Contact information for each appears in the supplier appendix.

America Online

America Online is generally considered to be the easiest to use, but it also can be slow because of the fast growth it has experienced—many nodes are simply overwhelmed. It offers its own software, free when you join, that gives you an attractive graphical interface that is easy to use. AOL offers access to most Internet services, in an easier-to-use format than many other providers. At this writing, full graphical Web access was on the way.

The content here places a heavy emphasis on recreation and news-related items. AOL offers access to many newspapers and magazines, including some of the most recent stories from the *Chicago Tribune, Business Week*, and *Spin*, among many others. Stock quotes are free of extra charges and are easy to access.

On the downside, there is no way to access forum messages automatically. Thus you are on-line longer, racking up connection charges.

CompuServe

CompuServe was the first nationwide on-line service, founded in 1979, focused heavily on financial and business information, but with over 900 interest forums, it is by no means limited to this. CompuServe is owned by H&R Block, and boasts the largest on-line access to computer support specialists—over 800 hardware and software companies maintain a presence here, answering user questions free of charge. You may have noticed that when you buy a new hardware or software package, it usually includes an introductory offer to connect you to CompuServe; this is why.

CompuServe offers its own graphical access software for Windows, Macintosh, OS/2, and other platforms, and a number of third-party offerings are available as well. Some of these third-party programs enable automated off-line access, which can help considerably at keeping your bill down. Internet access via CompuServe is pretty complete, and recently expanded to include graphical Web access, using their Spry Mosaic browser (you can also attach your own). Newspaper and magazine offerings are extensive, and include Associated Press stories right off the wire. Stock quotes are delayed by 15 minutes, and cost more after the first 20 you access per session.

The main downside of CompuServe is that it can get expensive, especially if you want to take advantage of its wide selection of business databases, press offerings, or financial resources. When you access most of these you get charged a surcharge, either per minute, per item accessed, or both, so costs can add up quickly. Also, there are charges for sending and receiving Internet e-mail messages.

Delphi

The attitude on this service is somewhat more friendly and easygoing than the first two. Delphi offers the most complete Internet access of these four at this time, with full (though currently textual) Web capability, as well as IRC and other Internet features the others don't offer. On the other hand, it offers fewer news or business database options than AOL or CompuServe.

This is definitely the place to go if you have an older computer; you'll find user forums focusing on older Apple machines, Commodore, and Amiga computers. The user interface is currently menu driven and some people find it hard to use, but a graphical version is promised soon (probably by the time you read this).

You won't find the same concentration of business-related information here as on CompuServe, but Delphi does offer stock quotes for a few cents per quote, and closing prices each day.

Prodigy

This service is a joint venture between IBM and Sears, Roebuck & Co. The first to offer a graphical Web browser, Prodigy gives you full Internet access in a very friendly package. It's lower on business content than the others, but higher on recreational and children's offerings. Stock quotes are available, as is support by some computer vendors, but surcharges apply. The Internet offerings include access to Usenet newsgroups and a graphical Web browser.

Prodigy currently does not offer a mechanism for automating retrieval of conference messages or sending e-mail, so on-line charges can add up fast. Some people find the interface inelegant; a major overhaul is planned by Fall 1995.

Make Sure They Go Where You Go

If you need to access your service from any medium-sized city in the United States or Canada, any of these services will suffice. Full worldwide access is more difficult. Of the four providers described above, only CompuServe offers access from cities all over the world. The alternative is to call into the service via a long-distance call, which can get very expensive. America Online plans to expand into Europe by late 1995, and others will very likely follow suit, so your options are likely to widen dramatically in the next year.

Even if you find a provider who offers service everywhere you need it, you are going to run into a host of other problems when you leave North America with your computer, including power and phone plug incompatibilities. Chapter 23 offers some suggestions for minimizing the difficulties.

Close to Home

Don't overlook the wealth of offerings that might be operating from a location right down the street. Though they might not offer nationwide coverage, local providers sometimes offer other advantages, including significant cost savings. In addition, their local focus means they are frequented by people who have interests that are likely to be similar to yours, and may have discussion groups on local restaurants, city politics, classes, user gatherings, and a more amiable culture than some of the national providers. Another benefit is the satisfaction of supporting your local economy instead of a big multinational corporation.

Like the big providers, many smaller BBS (bulletin board service) operators are branching out from their chat forums to providing Internet access. This access might be in the form of UNIX shell accounts or menu driven connections. More and more, they also offer SLIP or PPP connections, but the level of support will vary considerably; this is something to check into before you sign up.

Another thing to check into is how easy it is to dial into the service. Especially if it's new or growing quickly, some of these services can be heavily utilized, and it might be very hard to reach them during peak hours. Obviously your bargain access is no bargain if you can't reach them when you need to.

Another possible source of access, especially for UNIX shell accounts, might be your local university. Some offer shell accounts for very little cost to people in the local community, but the trade-off against lower cost is probably a much lower level of support.

Typical Pricing Structures and "Gotchas"

Pricing structures of the various providers change so often that it would make no sense to describe each structure here. Instead, below is a checklist of things to watch for when you are trying to compare providers.

Pricing Approaches

Each provider has developed a different approach to charging people for the time they use, and still making money on the service. Most have adopted some form of a flat rate coupled with an hourly rate when your usage goes over a certain number of hours per month, but here are some variations you might run across.

Flat rate per month. In this structure, you pay a fixed fee every month, and get unlimited access to the system. Expect the lowest level of customer support here, and therefore the most computer literate clientele.

Flat rate with time limits. Similar to the previous structure, but without unlimited access. The system might have timeouts that log you off after a certain amount of time, for instance, giving someone else a chance to dial in. Sometimes you can dial in again right away, sometimes you have to get in line with all the other users who want access.

Flat rate plus usage charges. This is the most common structure. Your monthly fee entitles you to a certain number of hours of access, after which you pay for the time on-line. Usually the on-line time is billed per hour in one-minute increments.

Charges for volume stored. Some systems charge users for the amount of data they store on the system hard disk, to keep the file system from getting clogged with user files. If you download files to the host system and then to your own computer, don't forget to delete them from the host disk to avoid this charge.

Charges for volume transferred. Similar to the structure described above, but here you get charged some fee for every kilobyte you move over the wire, whether you let it reside on the host system hard drive or not. This can get expensive quickly.

Flat fee plus network access charges. Especially on systems that you access via a toll free number, you pay per minute on-line. Ideally this charge is less than the comparable long-distance charges would be.

Charges based on speed of access. This variation is becoming less common, but at least one of the major providers described earlier in this chapter still uses it. If you access the service at 14.4 kbps, it can cost you twice as much as accessing at 2400 bps. If your provider does charge this way, save your high-speed settings for when you're doing data-intensive file transfers. For sessions that involve a lot of commands to the server and waiting for responses from the server, even in an automated session, you can save a lot of money by not paying for the high-speed access, since the responses slow things down so much. And if you are running a fully interactive session, such as a live chat discussion, don't use more than 300 bps—you can't type faster than that anyway.

Surcharges for certain services. Most providers who offer financial databases, article indexes, or airline ticketing services charge by the minute while you access the service, and possibly by the item referenced as well. This can add up quickly; know what you're going after before you log onto the service.

Charges for e-mail through gateways. If a principal reason for accessing the service is to be able to exchange e-mail with people on other networks, take a close look at the provider's e-mail pricing structure.

Charges for add-on e-mail services. On some systems, sending copies of messages to other recipients, or requesting notification when the recipient retrieves the message, can be subject to hefty fees.

Side Benefits or Drawbacks

It's important to compare on-line services not just on the basis of price, but also to consider the other benefits that might come with membership. Or to put it another way, if the service is very inexpensive, find out exactly what you are going to have to pay for yourself.

Support

Find out how much support they provide, how much it costs (if it does), how easy it is to reach the support technicians. Most providers will offer support through a newsgroup or forum; in some cases you may find live phone support.

Software

The survey of providers above presented a simple overview of the software that you get when you sign up for those services, but the offerings change frequently. Try to find out whether the software is free of charge, and whether they offer you help in getting it running. Make sure the software is available for your hardware platform, including the amount of memory you have and the version number of your operating system. Usually the software is free, or they charge you for the software but credit you the same amount in access charges.

Do they offer an automated installation for your TCP stack and Web browser, or must you find and assemble this software yourself? The higher tinker factor of the latter option is only for the rather dedicated, or for those who are willing to hire someone to get them set up.

On-Line vs. Off-Line Access

Some of the major providers support a mechanism that enables you to retrieve messages automatically, then disconnect from the service and read them at your leisure. While disconnected, you can review the messages and compose responses, then reconnect to the service to send your responses. If your service supports this type of access it can save you a tremendous amount of money in connect time charges. Some offer this feature for e-mail but not for forums or conferences. CompuServe has several third party software developers that offer and support programs that enable this; most of the other big providers do not currently support the feature. If your provider charges a flat rate for access, of course, the point is moot from the point of view of money, but is still worth thinking about out of consideration for other users trying to access the system.

Summary

You have a number of choices in getting access to the Internet or to an on-line service provider. The provider you choose depends on where you are, how much you travel, what services you expect to use, and how much you're willing to pay. This chapter described the typical pricing strategies, surveyed some of the major providers, and offered some tips on finding a provider closer to home.

chapter 22

Working Remotely

Introduction to Telecommuting

If you do most of your job on a computer, you may be able to do it from home. With advances in telephone technology, more powerful portable computing equipment, and better e-mail offerings, the home can be an extension of the office, or can be the only office. Many people who have made the shift to either partial or full-time working at home report that they are happier with their work, citing the increased flexibility and control over their personal lives and better ability to concentrate on tasks that require their attention.

Employers benefit from letting people work from home, too, not just because the workers are happier and more productive, but also because it can reduce their costs of maintaining office space for employees in some cases. Another factor that is leading companies to take a long look at letting employees telecommute is the environment. In the ten cities in the United States with the worst air pollution, according to the Clean Air Act of 1990, companies with more than 100 employees are required to reduce the amount of total time their employees spend on the road by 1996. In answer to this requirement, companies are looking at carpooling and other mass transit mechanisms to get their people to the office more efficiently, but at the same time they are also taking a long look at the options available to keep people from needing to come into the office at all.

After the Los Angeles earthquake in 1993, many people in the LA area used telecommuting for the first time. They couldn't drive to work on the ruined freeways, but they were able to log into their office LANs via phone lines and keep working that way. This solution focused the attention of much of the country on telecommuting in a new way.

Studies indicate that the number of American teleworkers is currently at 9.4 million, and that number is growing in the United States at an estimated rate of 20 percent per year. In Europe, the numbers are even higher; one in eight workers works from home at least some of the time.

Telecommuting takes several forms. Some people work at home full time; all the work they do is done from home. Other people use it frequently, but not constantly. These people might work three days a week in the office and two at home, or perhaps two weeks a month from home and another two weeks in the office (which might be across the country). On the most occasional end of the scale are the people who use their home offices to extend the workday or to give them an occasional break from the office environment.

This chapter focuses on the issues facing the people who spend more of their time in the home office than the corporate headquarters, but the concepts also apply to occasional home workers, as well as to working on the road. The next chapter focuses more specifically on the challenges of the portable office.

If you're considering spending more of your work week at home, the issues and strategies presented in this chapter will help you determine whether your work style is suited to the particular challenges, and what you need to consider to make it happen.

Making the Shift out of the Office

Moving from working in an office to spending at least some of your time working at home will require a number of changes. One of the first may be in the attitude of your supervisor—the first part of this section offers some tips on pitching the idea and fielding the objections. Having accomplished that, you may well also need to change your work habits when you get to your home office. The second part of this section explains how.

Talking the Boss into It

Despite all the great press that telecommuting is receiving right now, many managers are uncomfortable with the idea in practice, even those who are strongly committed to the concept. Some dislike the lack of control they feel when people are not in reach. Others have the nagging sense that if they can't see an employee working, he probably isn't.

When approaching your supervisor, you need to present the advantages from her point of view. The main issue to remember is that no boss will be interested unless she feels that it benefits not just you, but the bottom line. You need to show that your remote work helps meet the goals of the department and the company. Reassure her that you're serious and committed by writing up a formal proposal outlining your goals, schedule, and how you will make sure the work gets done.

The best way to start is to offer a trial period, perhaps a few days or once a week. Agree beforehand on some mechanism for measuring your productivity, and show that it goes up when you work at home.

If you're working on a contract basis, here's a tip from some professional consultants. Work through the first few months or first contract in their office so they get to know you, and have a chance to see how conscientious you are. Then the next time you're negotiating your fees, propose working some of the time out of the office. You might request a certain amount of money, which you may know they won't be able to meet. When they balk, offer to reduce the sum required if you can spend two or three days a week working from home. Many times, clients will be glad to find nonfinancial ways to compensate you, and everyone wins.

Moving out of Interrupt Mode

Having adjusted your supervisor and possibly co-workers to the change, the next step is to adjust your own work habits. Many offices are not terribly conducive to accomplishing work; there are phone calls, people dropping by, meetings, and other interruptions. In response, many people develop an interrupt-driven approach to work; they spend the day responding to interruptions instead of deciding what to do and focusing on meeting those goals. They get to the end of the day tired and having no tangible accomplishments, so they either stay late to catch up, take work home, or come into the office at dawn to try to get a jump on the day. This is not productive and is a big reason people make the shift to working at home, at least part of the time—they can better control the interruptions.

But they find that they don't know how. They spend their time checking e-mail, responding to the post, reacting to phone calls. Perhaps none of these things need to be urgently addressed right then, but the person is so used to responding to things as they come up that they find it difficult to work any other way.

In addition to the problems that come with loose discipline, the home environment has a whole different set of distractions from the workplace, many of which are just as compelling as the meetings and phone calls in the office. You walk into the kitchen for coffee and notice that the dishes need washing, or in the middle of writing a report one of the kids wants attention.

Since one of the advantages of home work was supposed to be more flexibility, you may not want to just close the door on the rest of the house and family for eight hours and act as if you're simply not there. But if you suspect you're spending more time on nonessential items, and find that you're not getting work done, consider reevaluating your approach.

One strategy to get a clearer picture of your home work is to make a parallel list of tasks every day. On the left, list everything you want to accomplish during the day; on the right, as you go through the day, list the accomplishments. These might well include personal and household accomplishments if that's where your time and values are. Refer to the list and update it frequently throughout the day. This will not only help you focus on what really needs to be done, it will also help you appreciate what you have accomplished that might not be so tangible. At the very least it will help you see where your time is going so you can adjust.

Financial Issues

More and more companies are embracing telecommuting because it makes people happier and therefore more productive, since with it employees have more flexibility and control over their lives and schedules. But there are financial forces at work here too. Until recently, employees have been the ones proposing telecommuting and managers have been reluctant. But as companies are casting about for ways to reduce costs, it's becoming something more companies, especially in Europe, are proposing to the employee instead. This is leading to a new set of questions. If the employee pays for a computer, fax machine, extra phone line, insurance, and electrical bill, not to mention house space, of a home office, the company benefits quite directly, and often the employee does not get compensated. In the United States, if you don't use your home office as your principal place of business, you don't even get to deduct the costs from your taxes. Consider the issues carefully, and talk to other home workers for suggestions when negotiating how much of the cost of setting up a home office the company will assist you with.

Working away from the Office

Since they can't see you and drop in to talk to you anymore, your co-workers and supervisor might be uncomfortable with the level of contact they have with you when you work from home. You may have to go overboard on keeping in touch in order to reassure your supervisor, at least at first. Be vigilant at answering e-mail promptly and thoroughly, and make sure the folks at the office can reach you easily by phone. After everyone gets used to the arrangement you may be able to relax the vigilance somewhat, but especially at the beginning, if they can't reach you, they are not likely to be happy with the experiment.

To some extent, the same rules apply after the trial period. Stay on top of the e-mail; don't let messages go unanswered any longer than necessary. It's useful to keep at least a few constant hours during which everyone always knows they can reach you—perhaps between 9 and 12 in the morning, for instance. Regular hours may be good for your own discipline and family situation, too. If you are going to be away from the home office during your usual scheduled time, leave a message to that effect so that co-workers don't have the sense that you have disappeared—a big complaint about remote workers.

Working out of the Home

People who choose to work at home on a permanent basis face many of the same issues as those who do it occasionally, and some extra issues as well. Here your home is your office, not just a temporary or infrequent extension of your office, so it may need more equipment and more complete isolation from the rest of the house than the less frequently used home office.

Organizational Issues

One challenge reported by many people who work out of the home is balancing the needs of the family and the business. Some advocate enforcing a strict division between work time and home time, just as if they had driven to work and were unavailable for the day. Just because daddy is in the next room does not mean he can talk to you right now. Similarly, they recommend a clear division between family space and work space; the office is strictly off limits to the rest of the family. This keeps office supplies from walking off, for instance.

Others suggest a more relaxed view without such strict separation, but still with a significant dose of discipline. All agree that being well organized is important, and that having someone to watch the children, at least part of the day, is essential. You can't run a business and be a nanny at the same time; they are both full time jobs.

Two other items should be subject to strict separation between work and personal life: the office space and the money. Keeping a separate office has the positive psychological benefit of enabling you to close the door on it at the end of the day, giving you an enforcement for "going home." And in the case of both the separate office and the separate money, there are significant tax reasons for making the divisions as well. In the United States, you simply can't deduct the costs of a home office unless it is a clearly separate room from the rest of the house (though this may change in the near future).

Personal Issues

The workplace provides more than a telephone line and a desk. It also offers contact with colleagues and a social environment that you don't have in a home office. For some people, the lack of this social interaction is a principal reason that they don't want to work from home.

To overcome the isolation, step up your contacts with other people. Join professional organizations in your area. Many on-line services such as CompuServe have forums that address these issues and provide contact to other people facing the same issues.

Working at home is not for everyone. The people who succeed at it and enjoy it the most tend to be self motivated and well organized, and they don't require the social network of the office. They must be disciplined, and able to resist lovely weather when a deadline is looming. They must be able to give themselves positive feedback, not expect a lot of it from others, and must be able to work without a lot of supervision.

Financial and Legal Issues

Working out of your home has a host of legal, tax, insurance, and liability issues that you might never have considered, and that don't face you when you just use your home as an occasional place to get some work done away from the corporate environment. Covering these issues goes far beyond the scope of this book, but there are several good books available that do address them. In particular, Paul and Sarah Edwards have written several books on the specific issues of working from home or running a business from your home. They also direct you to other resources for finding out how your local laws affect your home business.

The Partial Home Office

Called satellite offices in the United States, and teleworking centres in the United Kingdom, these businesses let you rent temporary space for your business. If you need to meet with a client, for instance, and your home is not an appropriate place for the meeting, one of these centers could be the answer. These centers provide not only meeting space, but also fax machines, phone lines, copiers, computers, secretarial services, and a quiet place to work. Many of them let you rent space by the hour, day, month, or year.

Taking advantage of the professional atmosphere offered by these services can help defuse one issue many home workers face: that there is a prejudice toward larger corporations. When you operate out of a corporate address such as this, even only a few hours a month, you seem bigger than you are. In some cases, this can be a significant advantage.

Office Mechanics

Every technology covered so far in the book helps you stay connected to clients or the main office more effectively. With an ISDN line you can make a remote connection to the office LAN that runs at comparable speeds to being actually connected to the LAN in the corporate headquarters. In addition, your co-workers can send you e-mail or notices just as easily as if you were right down the hall.

An IR-enabled laptop or notebook computer, as described in Chapter 15, can be a real boon. When you get to the office, it updates your workstation with all the changes you made the evening before, and you spend the day working on your workstation. When you're ready to leave, the software updates your laptop with the changes on your workstation so that the two are always synchronized.

Summary

Working away from the corporate office is gaining in popularity for a number of reasons, and can be an ideal solution to some of the vexing problems of the office environment. But it also comes with a number of problems of its own. Before making the change, you need to consider personal, social, financial, and organizational matters that affect remote work. This chapter covers the main issues involved in working in a fixed place away from the office; the next chapter discusses working in a mobile environment.

chapter 23

Taking the Show on the Road

Staying in Touch

One of the advantages of getting a portable computer is being able to carry it around with you. The ability to take your computer anywhere you might go has been promised for a long time, but the earliest portables weighed 20 pounds (10 kilograms) and had only a fraction of the power of a typical desktop system. As the weight has come down and the power has gone up, the promise of true portability is more and more real. The environment in which you operate the computer, though, is not the same everywhere you might go. Depending on where you go or what you want to do, you may encounter a variety of obstacles. This chapter describes them and offers some suggestions for getting around them.

Taking Your Computer with You

Whether you are traveling to Beijing, China, or Pittsburgh, Pennsylvania, you may well want to take your computer with you, for any number of reasons. You might want to record impressions of the trip, make presentations, or finish a report while away. Or you might take it as a tool to stay in touch with the office, friends, and family. As long as you prepare and plan ahead, this is quite possible. Here are some of the things you'll want to keep in mind before you head out the door.

Power Differences

If you stay within the United States and Canada, powering your computer won't present a problem. But once you move outside of this region, you may encounter power differences in three areas: voltage, frequency, and plug configuration. The world has not yet agreed on a single standard for these parameters, and doing so will be very expensive, since many countries will have to change all of their outlets, appliances, and transformers to conform to whatever standard is finally adopted. Some suggest that it will be many decades before you can take the same appliance with you without hassles. In the meantime, here's what you need to know.

Voltages

Voltages vary between countries. While the United States operates on 110 volts, many other countries use 220 volts. If you plug your trusty laptop into the wrong voltage, you could ruin it, plunge your hotel into darkness, or both. Most computer power supplies, though, either detect the voltage automatically (nearly all newer portables do) or have a switch on the back of the computer so you can change the voltage yourself.

The switch on a desktop system is usually labeled with something like "230/110," and is often located on the back of the computer near the socket where you plug in the power cord. You may need a screwdriver to switch it to the other setting; this is something you definitely don't want to be able to do easily, and never while the computer is running.

On your portable, look at the power supply for a label that says something along the lines of "AC110–240V." If you see this, you know your computer can handle voltages between 110 and 240 volts. If not, check with the manufacturer before you go.

Once you arrive and want to do some computing, check with the owner or manager of the building before you plug the machine in, especially if your machine does not automatically detect voltages. Every hotel owner in the world has a horror story about a tourist plugging in some electrical device and blowing the entire building's circuit—you don't want those stories to be about your $3,000 lifeline.

Cycles

The "AC" in the voltage description legend stands for *alternating current.* The rate at which the power alternates, or oscillates, varies from country to country. Problems with using the wrong frequency are especially noticeable on devices with motors, such as record players, sewing machines, and electric clocks, all of which could run at the wrong speed, but it might also affect your computer, especially the monitor, and the results there would be much more dire. Many monitors switch frequency automatically; check the documentation. The value will be either 50 or 60 Hz, and what your computer uses should be clearly marked.

Plugs

Once you have the voltage and frequency matched up, you still have to deal with the configuration of pins on the plug. In most places in the world you won't find the two flat blades used in the United States, so either take some plug adapters with you, or buy one when you get there (recommended only for industrial countries). Most computers have modular power plugs. With these, you plug a cord into a universal socket on the back of the power supply; the other end of the cord has a plug that fits the sockets of the local country. You can buy the cords in many airport shops or in larger electronic stores, especially in Western Europe. The airport shops offer convenience but high prices, but the electronics stores may be hard to find, so you'll probably want to go the latter route only if you are planning to stay a while.

General Power Issues

If you plan to work on the plane, take extra batteries. They are heavy but free you from frequent recharges, which would be impossible on the plane. Another advantage of having extra batteries along is that if you are in a particular country only a short time, you may not need to get a cord or plug adapter at all; just charge them all up before you go.

Notebook computer batteries tend to drain when you most need them, another reason to take extras. A small measure of defense against the problems this engenders is to get a notebook computer that has a meter that shows you how much power remains in the battery, rather than a system that just beeps when it gets low. With the meter you can plan ahead much better to save your work.

Companies that sell equipment to simplify traveling with your computer include Magellan's, Road Warrior, and TeleAdapt. You'll find contact information in the suppliers' appendix.

Phone Matters

Telephone plugs are even less standardized than power plugs. In Germany alone you might encounter four different types of phone plugs, as the country moves to update the phone system. The modular variety used in the United States is called an RJ11 plug.

The same companies listed above sell phone adapter plugs and may be able to advise you about what kinds of plugs are used at your destination. They can also supply kits that provide you with nearly every type of adapter used in different regions of the world, a worthwhile investment if you expect to take your computer to a lot of different places. If you are going to stay a while in the country, you might find it convenient to get a cord that has an RJ11 plug on one end and the local variation on the other, rather than stacking up piles of adapters.

Hot Phone Lines

In some ways, hooking into the phone line is less dangerous than getting power for your computer, because the plugs usually won't fit if they're the wrong type. But some phone systems run at higher voltages, and if you plug your PCMCIA modem into one of these, you could quickly fry the modem. This is not just an international issue; many U.S. hotels have digital switchboard systems that cause the problem. Get a line tester to be sure; IBM offers the tiny Modem Saver for a modest cost. In case the tester says the line is not safe for your PCMCIA modem, you'll probably want to have a backup pocket modem on hand.

Acoustic Couplers

If you're going a lot of different places and can't be sure if there are modular plugs at all, consider the modest acoustic coupler. Two decades ago, this device was the only mechanism for connecting to another computer over a phone line, and it is still the unit of choice for reporters, researchers, and others traveling to areas with less standardized phone systems, or where the system is not modular, or where they have to connect via pay telephones. The sources listed above should have these as well. They are not nearly as convenient, or lightweight, as modems, but they do free you from some of their problems. This is also the solution to modern problems in those countries that use different tones than your modem generates; not all are standard.

Frequent users of acoustic couplers note that on pay telephones in the United States, the maximum data rate you are likely to achieve is 2400 bps. But they suggest dialing into on-line services on the 14.4 kbps numbers if you have the option, because those higher-speed modems offer error correction.

Software Settings

If the connection doesn't go through right away, check the tone/pulse setting in your communications software; the solution could be that simple. Another possibility is that the dial tone frequency is one that your modem does not recognize. If you suspect this is the case, you may need to adjust the modem's initialization string in your communications software to disable the dial tone recognition feature. Take your modem documentation with you in case you run into this problem; for most Hayes-compatible modems, you just need to add ATX3 to the initialization string.

Computers and Airports

The first thing most new portable computer owners want to know is whether putting the computer through the baggage check X-ray machine will harm it or their diskettes. It will not; thousands of computers go through these machines every day, and work perfectly afterward with no loss of data. What you do want to avoid is getting the computer near the hand-held metal detector wands used in some airports to search passengers; these devices emit powerful magnetic fields that can damage disks. Your bag is usually going through the X-ray machine while you're being searched with the wand, though, so this is usually no problem. If you're still not comfortable with sending the computer through, though, you have the right in the United States to ask for hand inspection of your bag. The security personnel will probably ask you to turn the computer on to prove that it is not a bomb, so make sure the battery is charged up. In other countries, they may insist that they X-ray your bag, but it's nothing to worry about.

In fact, sending it through the X-ray machine has an advantage: you don't have to display the computer to the eyes of everyone who might walk by. As in any public place, it's better to keep a low profile with your computer, so as not to tantalize thieves. A later section in this chapter discusses security concerns.

Some devices, especially communications devices such as cellular phones and modems, do have the potential to interfere with the plane's circuitry. The jury is out on whether some other electrical devices interfere, but airlines would rather be safe than sorry, as would you and everyone else aboard. Thus they will ask you to turn it off during the critical takeoff and landing phases; since all of your belongings must be tucked away during that time anyway, this is no great hardship.

The Hotel Room Crawl

You've arrived at the hotel and are ready to check your e-mail. You discover that the phone jack is under the window, and the only electrical outlet is under the bed on the other side of the room, so you have to work halfway under the bed. To minimize this annoyance, take long cords for both the electricity and the modem. Most of the major hotel chains that cater to the business traveler in the United States have fixed this, and offer computer-ready rooms with power and phone jacks near a desk or table, but those long cords come in very handy if things have not yet been upgraded to this standard.

If your room doesn't offer these features, try asking at the desk about alternatives. It is sometimes possible to dismantle the phone jack and hand-wire a modem connection, and some travelers advocate all manner of experimenting with this. This book does not, for several reasons. First of all, there are too many combinations; there is no one way to hack the phone systems of the world. Second, tearing up the property of your hosts is unethical, at least. Third, too many times people not only don't put things back the way they found them, they ruin the telephone connection in the room. So persist if you must; but do be considerate and put things back the way you found them.

Note that getting connected is only half the battle. After you get a dial tone, you have to figure out how to dial out of the hotel's PBX system. Many hotel phones require you to dial a 9, 1, or 0 before the number, then wait while the system gets you an outside line. You can program the prefix and a delay into your dialing list, but this means either remembering to modify the number before each call or carrying two versions of your connection list, one for home and one for the road. Look for dialing software that makes this change easy.

Security Matters

Portable computers are very attractive to thieves, precisely because they're portable, expensive, and easy to resell. Losing the machine would be a double loss to you, of both property and data, so make a backup of at least the most critical data before you go, and look into trip insurance. Most policies that cover computers do so only as a special rider.

Try to use the machine as little as possible in very public or unsafe environments. Consider getting a bag for it that doesn't look like a computer bag. Some briefcase styles are fairly subtle, and it's possible to get daypacks equipped with the necessary padding to protect the computer. As with any valuables, your first line of defense is a strong dose of caution, backed up with a decent insurance policy.

On-Line Away from Home

Having leapt all the technical hurdles, how can you make sure you can still get any e-mail once you can dial out on your modem? Here are some tips.

E-mail Issues

When dialing in for your e-mail from abroad, the easiest way is to make a long-distance call with your modem back to a working number at home. This is also by far the most expensive option; first-minute long distance charges to the United States can run upward of $6. Instead, see if your network has a local node close to the city you're in. MCI Mail and CompuServe are your best bets here; both have fairly extensive worldwide networks of hubs, so even though you might not be able to dial in as a local call, you probably won't have to call very far across the planet to reach the nearest one. Take instructions on how to connect before you go; the mechanism is different in every country. Sometimes it's as simple as dialing a direct number that connects you to another modem; other times you have to dial into a packet network that requires a particular set of commands in order to connect (and these networks often assess per-minute surcharges on top of the cost of using the on-line service itself). Each of the services will be able to provide you with information on connecting from various destinations.

Within the United States, many on-line services offer toll-free 800 numbers, so that you don't have to carry a list of local phone numbers with you. When you call one of these, you usually get charged by the minute for use of the network; something to consider before attempting an interactive session or a long download.

Remote E-Mail Software Wish List

Your e-mail software can help or hinder your on-the-road message retrieval; here are points to consider when feature shopping.

Many remote versions of LAN e-mail systems don't allow you to fully connect to the LAN server. This means that when you connect, you get all of the unread messages in your In box, whether you want them or not. This can be expensive when you dial in long distance and someone has sent you a one-megabyte graphics file. Some systems let you query what messages are waiting, and select only the ones you want to download.

A difficult problem for e-mail software manufacturers is allowing remote users to delete the messages they no longer want; in short, to manage their mailboxes remotely. Most simply don't allow it; you can only delete messages from your master e-mail mailbox when you are connected to the LAN; the rest of the time you are only working with a copy of your messages. Some manufacturers are working on solutions to this problem; it's something to look for, but currently the most viable alternative is to simply dial into the office LAN to get your e-mail.

SUMMARY

Being able to take that superlight small computer with you does you little good unless you can use it when you get there. You need to consider power needs, phone system details, plug configurations, and network availability. The best way to be sure you have what you need is to talk to someone who has been to your destination and successfully connected a computer while there. Check in the travel section of your on-line service for tips before you go.

SECURITY AND THE REMOTE CONNECTION

chapter 24

GIVING AWAY THE FARM

There's a downside to all this connectivity—remote connections work two ways. Not only do you now have the means to call out, others have the means to call in. Keeping them out, and keeping your data secure, is a major part of managing remote connections, and it has many different aspects. What you need to do depends on the type of remote access you are using.

This chapter describes what hackers are doing to get into systems and what types of protection you can use to keep them out, then advises you on what to use based on your usage situation.

The word *hacker* to describe someone who maliciously breaks into a computer is used here with some regret. A few years ago, a hacker was someone who took passionate delight in finding out the details of how every aspect of computers worked. The word was used to distinguish a hacker from a *cracker,* who was breaking into systems illegally, and possibly damaging them. Due partly to misuse in the mainstream media, though, the former word has come to describe the latter creature, and this chapter reflects that changed usage.

Before you become alarmed at the dire risks described in this chapter, keep in mind that the level of risk to your data is directly proportional to the openness of your system. If you only turn on your modem to dial out to check your Prodigy e-mail a few times a week, you run very little risk of being on the wrong end of the problems described in this chapter. If you operate a public bulletin board system, on the other hand, or manage a LAN into which the local population can call at any time, you need to pay much more attention to the dangers and defenses described in this chapter.

The Risks

This section starts the discussion of security with an overview of what's at stake. Some of the risks are obvious, others more subtle. You'll want to consider the level of risk when you try to determine what security measures are appropriate for your installation. In effect, you'll be stating how big a risk you are willing to take.

Loss of Data

An intruder could log into your computer and download the entire contents of your hard disk (or disks). While this is not precisely a loss, because you still have the data, it is less valuable to you because now someone else has it too. This pertains especially to sensitive or competitive information in a business—a rival stealing your sales plan and price list, for instance, could be a severe blow if you're both competing for the same customers.

Corruption or Destruction of Data

Most hackers are more interested in finding out how to get into a system than in causing trouble there, but malicious activity does happen. You might not mind that someone has access to your e-mail correspondence with your mother, but you might not want them to delete it all.

Violation of Privacy

This type of damage is difficult to quantify, and can run the gamut from your love letters being posted on the office bulletin board (using the office e-mail system for personal messages is not a good idea!) to having your credit card number stolen and used to charge enormous amounts of money to your account. Anything that is stored on a computer anywhere is at risk, and some of it you can't protect yourself against. You might not want an unauthorized intruder to be able to download your entire credit history, but unless you manage the credit record companies' computers, you can't do much about it. You can, however, protect some aspects of your private data, especially that which resides on your own systems.

Viruses and Other Destructive Software

A virus is a program that loads itself onto your computer when you run an infected program or load an infected diskette. Some are relatively harmless, simply displaying clever messages on your screen at random times, for instance. Most of them, though, are malicious, designed to damage data, perhaps after a time period of months after you load them.

The chance of acquiring a virus is highest for those people who use remote access to get files and programs from other computers. If you don't know the source of the program, it could have a virus, and that virus could wipe out some or all of your system. Even if you do know the source you should check the software for viruses; your friend's computer might have become infected by accident too.

Protecting Yourself

All of this may sound alarming, and there are network administrators who simply disconnect all modems in the building in response. You don't need to do anything so drastic. Here are a number of things you can do to protect against these intrusions, some of them remarkably simple.

Passwords

The first line of attack of most intruders is to run software that automates the login process to your computer, running through the most common or likely passwords and trying them all in turn until the program finds one that works. Careful password selection, though, can go a long way toward foiling this type of attack.

An enormous number of passwords—an estimated 80 percent—are simply words found in the dictionary, and intruders use this fact to break into many accounts quickly. The first thing their automated software is likely to do is cycle through words and names.

Thus the lowly password is more important than you might think. Avoid using passwords that are made up of meaningful words, including names. The second thing an intruder will try, if he knows anything about you, is information that is important to you, such as your birthday, that of your kids or spouse, or part of your phone number. These are all very common passwords, and very easy to guess.

Further password selection tips: Mix the case of the letters, rather than using an all uppercase or all lowercase password. Use numbers and punctuation marks, not just letters. If your system allows it, use some control characters; this will increase the security of the password dramatically. Finally, change the password often.

Now that you, and perhaps all the users of your LAN, have these wonderfully secure new passwords, how are you going to remember them? Making them harder to guess creates another problem, that it makes them harder to remember. If you do write the password down somewhere, make sure you don't do it on a piece of paper taped to your monitor. Protect the password the way you would a PIN number for a cash debit card.

Record Keeping

If you administer a system that allows the outside world to log into it, you'll want to install software that reports suspicious access. At the very least, install software that logs errors. When you examine the log and see that a particular user mistyped his password 500 times in the span of a minute, you know someone is trying to gain unauthorized access to your system. The next level of protection would be software that permits users to mistype their passwords only a limited number of times, perhaps five, before locking the account and informing them that they must call the system administrator.

Another useful type of record keeping keeps track of access times. When the log shows that Marcia was logged on between 3 and 5 AM, though the building access records show that she didn't come in until 8:30 that day, you know you have a problem.

Encryption

Whenever you send sensitive information over a nonsecure link, you should consider encrypting it. This goes for sending your credit card number over the Internet or holding a company strategy video conference over an unsecured ISDN link. Many people routinely encode all of their correspondence. Recent years have brought encryption algorithms that cannot be broken with current computing power. Such algorithms, called *public key encryption algorithms*, offer excellent security and are easy to use. A popular software package in the United States that offers this type of encryption is called PGP, and is easy to find at many sites on the major on-line services and the Internet.

To secure lines for such things as video conferences, consult with the manufacturer of the equipment or the provider of the line access to find out what options they offer to protect the data. DES is the algorithm most commonly used in this environment, but is by no means the only one.

Encryption is a huge topic, and is complicated by issues relating to personal privacy, national security, and criminal activity. The controversy arising out of all of these issues may affect how easy it is for you to get encryption software for personal or professional use. For an evenhanded overview of the issues, consult the article *Crypto Policy Perspectives* (Landau et al.) in the August 1994 issue of *Communications of the ACM*. For an excellent treatment of the algorithms, get a copy of Bruce Schneier's book *Applied Cryptography: Protocols, Algorithms, and Source Code in C* (John Wiley & Sons, 1994).

Filtering

This defense is relevant only if you have systems available to the general public, especially via the Internet. Packet filtering takes advantage of the knowledge you have about what packets coming into your network should contain, and ignores those that don't match.

From the other side, people who try to break into your machine may try to use a technique called *address spoofing*. They create network packets that appear, at least from the header information, to be valid network packets. Once the packets are admitted to your system, the instructions contained in them can be used to simulate legal sessions, download files from your computer, log into your computer, and launch other attacks using your computer as a base. Packet filters prevent this by raising the knowledge in the network about what kinds of packets are legal. Many commercial vendors provide packet filter packages, especially for TCP/IP networks.

Another type of filtering is *application filtering*. In this mode, your system simply prevents certain types of applications from running at all. The powerful and easy to use MIME e-mail standard, for instance, automatically launches applications based on the description of the document encoded in the e-mail message. The idea behind this scheme is to simplify and standardize the complicated realm of e-mail attachments, by freeing the user from knowing what application she must use to view a particular attachment. If it's a graphic, the e-mail client launches a graphics viewer; if it's a word processing document, it launches the appropriate word processor. Malicious users have taken advantage of this mechanism, though, to launch other programs, including some that let them log into the system. In defense, a network administrator can prevent these applications from running, or require authentication.

Firewalls

Setting up a firewall represents the big guns of network defense. It is also the most popular way of protecting your Internet site. Usually running on a dedicated computer system, a firewall sits between your internal, protected network and the wild outside, allows only specifically allowed traffic to get to the internal network, and acts as a secure entryway for traffic coming from the Internet. Security mechanisms on firewall systems might combine all of the above protections.

Packet filters running on your firewall machine can protect you against a determined intruder who knows how to simulate valid network packets. A well-written packet filter can detect and discard bogus packets, but writing the packet filter correctly is something of an art. Consult *Firewalls and Internet Security*, by Cheswick and Bellovin, for a complete treatment of the issues and techniques involved in setting up firewalls. This book is useful to raise your awareness of any type of network protection; you need not have a UNIX system or an Internet-connected network to find it useful.

Firewall construction depends on your philosophy and budget. They are difficult to configure correctly and require constant attention, and to be maximally effective they require a dedicated machine. If you're not sure whether you need one, weigh the potential for loss against the costs carefully.

Finding the Security Solution for You

Not all of these security methods are suitable for everyone in every situation. This section describes typical remote access implementations, and what types of security you should consider in each case.

Security and Modem Surfing

If your remote access forays involve using your modem to visit other computers, and you have not configured your computer to be a host to allow others to dial into it, then your main security risks involve contracting viruses from software you download from remote systems. Get virus-checking software and run it on every program you download. While virus checkers are available from remote sites, there have been stories recently of the checkers themselves containing viruses. Instead, make the investment in purchasing a virus checker from a reputable software vendor.

If you ever send credit card information over your modem, be sure to encrypt the message before you send it. If this is not feasible (if the people on the other side don't have the same encryption software, for instance), phone them with the information. Credit card fraud is rampant and growing.

Security and the Dial-In Account

To protect your account on America Online or your local service provider, change your password often. If you send sensitive information from your account, send it encrypted. Note that even though an intruder might not be interested in what you have stored in your account area, he could use your account to launch further attacks, using you as a shield.

Security and Cellular Connections

As some people have learned rather painfully, and as you are probably aware, cellular and cordless phones are just radio transmitters, and anyone with a receiver can listen to your call. The same goes for any data you send over a cellular link, perhaps with your cellular modem. When you purchase one, ask about scrambling and encryption options.

Security and the Network Administrator

If your LAN resides only in your building and neither the server nor any of the client systems have modems or connections beyond your walls, you have no security problems because you have no remote access. Better stated, your security problems relate to internal staff walking out with diskettes full of sensitive data, or corrupting things from inside.

As soon as you add a link to the outside world, you have an access security risk. This includes providing a modem line to let an employee work from home. The paranoid response is to disconnect all modem lines, but this is not necessary. You just need to educate yourself on how to keep them safe.

Get all users to change passwords often, and back up the policy with software that forces them to change their passwords every few weeks. Educate them on choosing good ones, and explain why this is important. Some password programs prevent really stupid passwords, such as one or two character passwords or passwords that can be found in the dictionary.

Your next line of defense is to install a firewall, which lets you restrict access to just that one machine and free the users from having to know about the security matters. The books mentioned above are excellent resources if you need to go this route.

Not to Decide is to Decide

You can ignore the suggestions listed above, or delay choosing a method until some time in the future. But keep in mind that not doing anything is also a security method, and one that has a high possibility of running you into trouble. Even if you don't think you have much of anything on your machine for intruders to steal, there are still hackers out there who are not only after information, but who are malicious and would damage your system if they could get in. If you deplore high-security, paranoid approaches such as this, you're not alone. But there are in fact malicious people out there, though they are very few, and if you have data to protect you would be much better off doing so than hoping that you never run into any of them.

Summary

Computer security is a complex topic, and this chapter only skimmed the motivations, risks, and options. With the information in this chapter, you are now able to make an informed decision about the types of security you need, find out how to implement them, and understand what you're protecting yourself against.

appendix A

GLOSSARY

account

An administrative convenience for the system administrators of large systems to restrict access to the system to those whom they know are authorized.

account ID

The identifier with which you log into a system.

analog signal

A signal that varies in a continuous pattern, rather than in the discrete pattern of a digital signal.

API

Application Programming Interface. Usually a set of rules that define how two different software components should communicate.

Archie

Internet catalog utility that provides indexing to some of the thousands of files on the Internet.

asynchronous communications

Communications mechanism for sending data between two devices, in which the timing of the sending of each unit of data is not related to the timing of the other units.

bandwidth
The capacity of a transmission medium. Generally measured in bps, kbps, Mbps, etc.

baud rate
Measurement of speed of signaling, formerly used for modems.

Bitnet
Network of independent computer operators. Bitnet offers e-mail, file transfers, and other services. Grassroots and worldwide.

BRI
Basic Rate ISDN. This is the ISDN variety with two B channels and a D channel, the kind you're most likely to get at home. See also PRI.

CCITT
Old name for ITU-T, the international communications standard definition body.

cellular
Usually refers to the radio telephone network that uses a cell model for providing telephone service.

character recognition
Technology that examines an image of a sample of text and translates the image into a text representation of the document.

client computer
Any computer when it is acting as a satellite of a server or servers, requesting services from that server.

compression, data compression
Technique that examines data for redundant sections and stores or transmits only the parts that are unique, along with information about what was left out. At the other end, the decompression engine restores the data to its exact original configuration.

connection equipment
Modem of any variety, whether landline, cellular, radio, or satellite.

copper media
Phone lines made of copper wire, the most common type.

dedicated line

A phone line between two sites that is always available for those sites and for no others.

digital signal

One that varies in a discrete, on-and-off, pattern. Opposite of analog signal.

download

The act of transferring data from any other computer to yours. In a data transfer, you are always in the well. See also upload.

e-mail, electronic mail

Refers to the software and systems that let users exchange electronic messages with each other, or to one of those messages itself.

encryption

Technology designed to make data unintelligible to anyone but those with the authorized keys.

error correcting software

Software used especially for data transmission that detects transmission errors and can correct them, often by arranging to have the erroneous data re-sent.

fax/modem

A modem with built-in fax capabilities.

fiber optic lines

Phone lines made of glass fibers. These can transmit faster, and therefore more, than copper lines.

forum

An electronic discussion group or bulletin board, where members post messages on various topics, to which other members respond.

FTP

File Transfer Protocol, the Internet utility for transferring files between computers.

Gopher

An Internet utility and protocol that provides a common, menu-driven interface to many of the Internet's offerings.

GSM

The digital cellular network that operates at 900 MHz in Europe.

handwriting recognition software

Software that attempts to "read" human handwriting and translate it into computer readable text. Not yet ready for wide-scale adoption.

Hayes command set

Set of standard commands that enable computers and modems to communicate in an expected fashion.

host computer

The computer that is providing services to other computers, which are then the clients.

IMAP

Internet Mail Access Protocol. E-mail protocol used on Internet servers, more advanced than but compatible with POP.

infrared

High-frequency band of the radio frequency spectrum. Also refers to technologies that use this frequency for short-range wireless communications.

interactive connection

A link to another computer during which you type in your commands "live," rather than prescribing them before you log in (an automated connection). Can also refer to connections when you are communicating with another human being via the computer.

Internet

The network of networks, based on a U.S. defense and educational network but now worldwide, that is also sometimes called the Information Superhighway.

internet

Any group of interconnected networks.

interrupt request line

See *IRQ*.

IR

An abbreviation for infrared.

IrDA

A standards committee that defines how infrared devices should communicate. Also the name of the standard that they have devised.

IRQ

Interrupt request line. One of the two parameters (with a memory address) that together specify a port on a PC.

ISDN

Integrated services digital network. Relatively high bandwidth protocol for digital communications over single phone lines. Requires new phone company switches and new equipment between the wall and your computer, but greatly increases your communications capabilities.

ITU-T

Formerly called the CCITT, this is the standards body under the UN that defines the technical details of how international telecommunications systems should interact.

LAN

Local area network. A set of computers and the equipment that links them together within a single building or other fairly narrow region.

landline modem

Conventional modem that runs over phone lines. This term is usually used to distinguish conventional modems from their wireless (cellular or radio) cousins.

legacy system

Any system that you have already invested money in. One of the biggest challenges of getting new technology is usually ensuring that it works with your legacy system.

link

The connection between two computers, or sometimes, the time during which they are connected.

local computer

The one sitting in front of you. The other one, the remote computer, might be across the room or across the planet.

logging in

The act of establishing a connection with a remote computer and identifying yourself to that system, usually by typing in your account ID and password.

logging out

The act of terminating the session and connection with the remote computer.

logon ID

See *account ID.* This term is more frequently used in LAN environments, while the same information is often called an account ID when you're dealing with a service provider.

machine

Another term for a computer.

mail robot

Automated software that can examine each e-mail message as it comes in and determine what to do with it based on certain rules that you specify.

mailing list

An e-mail based discussion mechanism or information distribution mechanism. Subscribers all receive the list message when it comes out, and in the discussion format, can send messages addressing issues in the previous message; these comments also get sent to everyone on the list. In the information format, the sending goes out only; comments are not allowed.

mainframe

A large and expensive computer that can handle thousands of simultaneous users. Most mainframes are made by IBM.

medium

Whatever the connection operates over. Could be phone lines, ISDN lines, radio waves, cellular waves, or satellite connections.

MIME

Multipurpose Internet mail extension. A standard for sending e-mail attachments of various kinds, from word processing documents to multimedia movies. Defines how the documents should be sent and how the receiver can recognize them.

modem

Modulator/demodulator, the device that translates your computer's digital signals to analog to send over a phone line, and then translates the incoming analog signals back into your computer's digital language.

network

Any set of connected computers and the equipment and media that connect them.

network providers

Usually refers to the providers of digital lines. May be, but is not necessarily, your local telephone company.

newsgroup

The Internet equivalent of a forum.

OSI

Open Systems Interconnection. This is a protocol model defined by an international standards group that defines how the various elements of networking share the duties of getting data from one place to another.

packet

A single collection of bits that is organized to be sent over a connection.

parity bit

A bit in a data word, used in telecommunications to indicate the type of error checking that is in use.

PC

Abbreviation for personal computer. Can mean any personal-sized computer, as opposed to micro or mainframe computers, and so can refer to Macintoshes, Ataris, and others. Also a shortened form of IBM-PC, in which case it only refers to those Intel 80xxx-based computers that run DOS and Windows.

PC Card

The new and more mnemonic name for the PCMCIA standard. Could cause confusion with a casual term for integrated circuit boards that fit inside computers, which are sometimes also called cards.

PCMCIA

The name of a standard that defines how these extra-small devices work, it also refers to the form factor of the device itself. Comes in Type 1 (the smallest), Type 2, Type 3, and Type 4.

PDA

Personal digital assistant. Ultra-portable pocket computers designed to take the place of checkbooks, address books, planning calendars, notepads, and appointment books.

POP

Post Office Protocol, a messaging protocol widely used on Internet e-mail servers. Also, Point of Presence, a term for the nearest telephone network connection.

port

A channel between your peripheral device, such as a modem, and your computer processor.

PRI

Primary rate ISDN. This variety has much more bandwidth than BRI, with 23 B channels and a D channel. Also known as T1.

protocol

Any standardized set of rules that specifies how data communication should proceed over a connection.

remote access

The act of connecting to another computer that is not right in front of you in order to work on it, access its files, or communicate with others.

RFC-822

The name of the standards document that defines how Internet e-mail is sent and received.

rich text

Text that includes formatting information such as fonts, colors, or other details.

scanner

A device that takes a digital "picture" of a document and sends it to a computer. Especially useful when used in conjunction with character recognition software, in which case you can input existing documents into your computer without having to type them in by hand.

sysop

System operator. The person who runs a dial-up computer bulletin board system or who moderates an on-line service forum.

system operator

See *sysop*.

TCP/IP

A protocol suite developed by the U.S. military research branch, and upon which the Internet is based.

telecommuting

The act of using your computer and telecommunications technology to work somewhere other than the office, usually at home.

terminal adapter

A device that connects between your computer and your ISDN line, translating the data on the line to a format for your computer.

terminal emulation

Technology that enables your PC to simulate the commands and interface of a mainframe terminal, so that you can use your PC to connect to a mainframe. Usually comes with communications software packages.

third party

Anyone other than you and your main provider. If you buy a new PC and it has a communications software program already installed, but you don't like it and go buy something else at the software store, you have purchased a third-party solution.

throughput

The amount of data transmitted at a particular point in time. This is the actual achieved, as opposed to the bandwidth, which measures the potential.

transmission speed

Speed at which your communications device can send and receive data. Typical speeds for modems are 9600 bps, 14.4 kbps, and 28.8 kbps.

Trojan horse

A program that looks innocuous and performs a useful function, but that also contains destructive software that invades your system through the useful utility.

upload

The act of transferring files or data from your computer to another.

Usenet

The collection of Internet discussion groups known as newsgroups. Can be addictive.

user ID

See *account ID, login ID.*

UUENCODE, UUDECODE

Pair of programs that translate binary files into ASCII (UUENCODE) and back to binary (UUDECODE). Necessary when trying to send binary files through e-mail gateways that do not permit binary transfers.

Veronica

An Internet search tool. You supply a string of search words and Veronica searches all of the Gopher menus it can find for a match.

virus

Software designed to invade a system, self replicate, and from there invade other systems. Often but not necessarily malicious and destructive.

WAIS

Wide area information service. Powerful Internet search tool that performs full text indexed searches of Internet data repositories.

WAN

Wide area network. A network connected over a larger geographical area than a LAN, and so requiring different technology to connect. Could be adjacent buildings or different continents.

Web browser

The client software needed to access the World Wide Web. Netscape is one of the most popular.

wireless connection

Any connection that operates over radio frequencies, whether cellular, radio, infrared, or satellite.

World Wide Web

Information storage and retrieval system that uses hyperlinks to connect the various computers in the web.

worm

Similar to a virus but uses a different mechanism to replicate.

X.400

The ITU-T e-mail standard, aka MHS. Describes how messages must be formatted, addressed, and exchanged.

appendix

B

VENDOR LIST

CONTACTING THE SUPPLIERS

This appendix contains contact information for all of the suppliers and service providers listed in the book. Where possible, it includes local phone numbers as well as 800 numbers, for those of you outside of the 800 service. If available, it also includes a fax number or an e-mail address.

SERVICE AND INFORMATION PROVIDERS

America Online, Inc. 800-827-6364 or 703-893-6288

ARDIS. 708-913-4233

CompuServe Information Services. 800-848-8199 or 614-457-8650

Delphi. 800-695-4005. e-mail: info@delphi.com

Dialog Information Services. 800-334-2564 or fax 415-858-7069

Dow Jones News and Retrieval Service. 800-522-3567 or 609-452-1511

MCI: MCI Mail. 800-444-6245 or 202-833-8484

Mead Data Central: Lexis/Nexis. 800-227-4908 or 513-865-6800

Prodigy. 800-776-3449 (not available outside the United States)

RadioMail Corporation. 800-597-6245 or 415-286-7800.
e-mail: info@radiomail.net

RAM Mobile Data. 908-602-5500

Ziff-Davis: ZiffNet. 617-252-5566

Hardware Vendors and Integrated Solutions

Apple: Macintosh, Newton Message Pad. 800-538-9686 or 408-996-1010

Digital Communications Associates: IRMA board. 404-740-0300

Hewlett-Packard: OmniBook PDA. 800-443-1254

Motorola: Marco and Envoy PDAs. 800-894-7353

Socket Communications: PageCard Wireless Messaging System. 800-552-3300.

Communications, Messaging, and Networking Software

Datastorm Technologies: ProComm Plus. 800-315-3282 or 314-443-3282

General Magic: MagicCap PDA operating system. 415-965-0400

IBM Personal Computer Company. 914-766-3000

Intel: ProShare conferencing software. 408-765-8080

Lotus: Notes, cc:Mail. 415-961-8800, fax 415-961-8400

MESA Technologies: Conference+. 617-964-7400, fax 617-964-4240

Microcom, Inc.: CarbonCopy. 800-822-8224 or 617-551-1000, fax 617-551-1021

Microsoft: Windows, Word, Word for Windows, Windows NT RAS, Microsoft Mail. 206-882-8080

Novell: NetWare, NetWare Connect, WordPerfect, GroupWise, Perfect Office. 800-638-9273 or 801-429-7000

Shiva: LANRover. 800-458-3550 or 617-270-8300

Symantec: pcAnywhere. 800-441-7234 or 408-253-9600

Traveling Software: LapLink. 800-343-8080 or 206-483-8088

Triton Technologies: CoSession. 800-322-9440 or 908-855-9440, fax 908-855-9608

Utilities

Caere Corp.: PageKeeper. 800-535-7226 or 408-395-7000

Delrina: WinFax PRO. 800-268-6082 or 408-363-2345

Inset Systems: Hijack PRO Graphics software. 800-374-6738

Intuit: Quicken and Pocket Quicken finance software. 800-924-1040

PkZip compression software for PCs. 414-354-8699, fax 414-354-8559

Aladdin Systems: StuffIt compression software for the Macintosh. 408-761-6200

SystemSoft Corp.: CardWizard Pro. 508-870-0050, fax 508-651-8188

Equipment and Support for Mobile Computing

Magellan's travel catalog. 800-962-4943

Road Warrior. 800-274-4277, e-mail: questions@warrier.com

TeleAdapt. 408-370-5105, fax 408-370 5110, e-mail 72623.706@compuserve.com

INDEX